STOCHASTIC DYNAMICS

Modeling Solute Transport in Porous Media

NORTH-HOLLAND SERIES IN

APPLIED MATHEMATICS AND MECHANICS

VOLUME 44

ELSEVIER
AMSTERDAM – BOSTON – LONDON – NEW YORK – OXFORD – PARIS
SAN DIEGO – SAN FRANCISCO – SINGAPORE – SYDNEY – TOKYO

STOCHASTIC DYNAMICS

Modeling Solute Transport in Porous Media

DON KULASIRI and WYNAND VERWOERD

Centre for Advanced Computational Solutions (C-fACS), Lincoln University, Canterbury, New Zealand

N·H

2002

ELSEVIER

AMSTERDAM – BOSTON – LONDON – NEW YORK – OXFORD – PARIS
SAN DIEGO – SAN FRANCISCO – SINGAPORE – SYDNEY – TOKYO

Elsevier
Radarweg 29, PO Box 211, 1000 AE Amsterdam, The Netherlands
The Boulevard, Langford Lane, Kidlington, Oxford OX5 1GB, UK

First edition 2002

British Library Cataloguing in Publication Data
Kulasiri, Don
Stochastic dynamics : modeling solute transport in porous media. - (North-Holland series in applied mathematics and mechanics ; 44)
1.Stochastic processes 2.Differentiable dynamical systems 3.Porous materials - Mathematical models 4.Solution (Chemistry) - Mathematical models
I.Title II.Verwoerd, Wynand
519.2'3

ISBN 0444511024

Library of Congress Cataloging in Publication Data
Kulasiri, Don.
Stochastic dynamics : modeling solute transport in porous media / Don Kulasiri and Wynand Verwoerd.
p. cm. -- (North-Holland series in applied mathematics and mechanics ; v. 44)
Includes bibliographical references and index.
ISBN 0-444-51102-4 (hb. : alk.paper)
1. Porous materials--Permeability--Mathematical models. 2. Transport theory--Mathematical models. 3. Fluid dynamics--Mathematical models. 4. Stochastic processes. I. Verwoerd, Wynand S. II. Title. III. Series.

QC173.4.P67 K85 2002
620.1'16--dc21

2002032218

ISBN: 0-444-51102-4
ISSN: 0167-5931 (Series)

Transferred to digital printing 2007

To my wife Sandhya for her support, encouragement and love.

Don Kulasiri

To my wife Nona and our children with love.

Wynand Verwoerd

Preface

We have attempted to explain the concepts which have been used and developed to model the stochastic dynamics of natural and biological systems. While the theory of stochastic differential equations and stochastic processes provide an attractive framework with an intuitive appeal to many problems with naturally induced variations, the solutions to such models are an active area of research, which is in its infancy. Therefore, this book should provide a large number of areas to research further. We also tried to explain the ideas in an intuitive and descriptive manner without being mathematically rigorous. Hopefully this will help the understanding of the concepts discussed here.

This book is intended for the scientists, engineers and research students who are interested in pursuing a stochastic dynamical approach in modeling natural and biological systems. Often in similar books explaining the applications of stochastic processes and differential equations, rigorous mathematical approaches have been taken without emphasizing the concepts in an intuitive manner. We attempt to present some of the concepts encountered in the theory of stochastic differential equations within the context of the problem of modeling solute transport in porous media. We believe that the problem of modeling transport processes in porous media is a natural setting to discuss applications of stochastic dynamics. We hope that the engineering and science students and researchers would be interested in this promising area of mathematics as well as in the problems we try to discuss here.

We explain the research problems associated with solute flow in porous media in Chapter 1 and we have argued for more sophisticated mathematical and computational frameworks for the problems encountered in natural systems with the presence of system noise. In Chapter 2, we introduce stochastic calculus in a relatively simple setting, and we illustrate the behavior of stochastic models through computer simulation in Chapter 3. Chapter 4 is devoted to a limited number of methods for solving stochastic differential equations. In Chapter 5, we discuss the potential theory as applied to stochastic systems and Chapter 6 is devoted to the discussion of modeling of fluid velocity as a fundamental stochastic variable. We apply potential theory to model solute dispersion in Chapter 7 in an attempt to model the effects of velocity variations on the downstream probability distributions of concentration plumes. In Chapter 8 we develop a mathematical and computational framework to model solute transport in saturated porous media without resorting to the Fickian type assumptions as in the advection-dispersion equation. The behavior of this model is explored using the computational experiments and experimental data to a limited extent. In Chapter 9, we introduce an efficient method to solve the eigenvalue problem associated with the modeling framework when the correlation length is variable. A stochastic inverse method that could be useful to estimate parameters in stochastic partial differential equations is described in Chapter 10. Reader should find many directions to explore further, and we have included a reasonable number of references at the end.

We are thankful to many colleagues at Lincoln University, Canterbury, New Zealand who encouraged and facilitated this work. Among them are John Bright, Vince Bidwell and Fuly Wong at Lincoln Environmental and Sandhya Samarasinghe at Natural Resources Engineering Group. Channa Rajanayake, a PhD student at Lincoln University, helped the first author in conducting computational experiments and in implementation of the routines for the inverse methods. We gratefully acknowledge his contribution.

We also acknowledge the support given by the Foundation for Research, Science and Technology (FoRST) in New Zealand.

Don Kulasiri
Wynand Verwoerd
Centre for Advanced Computational Solutions (C-fACS)
Lincoln University
New Zealand

Contents

Chapter 1

Modeling Solute Transport in Porous Media

1.1 Introduction

The study of solute transport in porous media is important for many environmental, industrial and biological problems. Contamination of groundwater, diffusion of tracer particles in cellular bodies, underground oil flow in the petroleum industry and blood flow through capillaries are a few relevant instances where a good understanding of transport in porous media is important. Most of natural and biological phenomena such as solute transport in porous media exhibit variability which can not be modeled by using deterministic approaches, therefore we need more sophisticated concepts and theories to capture the complexity of system behavior. We believe that the recent developments in stochastic calculus along with stochastic partial differential equations would provide a basis to model natural and biological systems in a comprehensive manner. Most of the systems contain variables that can be modeled by the laws of thermodynamics and mechanics, and relevant scientific knowledge can be used to develop inter-relationships among the variables. However, in many instances, the natural and biological systems modeled this way do not adequately represent the variability that is observed in the systems' natural settings. The idea of describing the variability as an integral part of systems dynamics is not new, and the methods such as Monte Carlo simulations have been used for decades. However there is evidence in natural phenomena to suggest that some of the observations can not be explained by using the models which give deterministic solutions, i.e. for the given sets of inputs and parameters we only see a single set of output values. The complexity in nature can not be understood through such deterministic descriptions in its entirety even though one can obtain qualitative understanding of complex phenomena by using them. We believe that new approaches should be developed to incorporate both the scientific laws and interdependence of system components in a manner to include the "noise" within the system. The term "noise" needs further explaining.

We usually define "noise" of a system in relation to the observations of the variables within the system, and we assume that the noise of the variable considered is superimposed on a more cleaner signal, i.e. a smoother set of observations. This observed "noise" is an outcome of the errors in the observations, inherent variability of the system, and the scale of the system we try to model. If our model is a perfect one for the scale chosen, then the "noise" reflects the measurement errors and the scale effects. In developing models for the engineering systems, such as an electrical circuit, we can consider "noise" to be measurement errors because we can design the circuit fairly accurately so that the equations governing the system behavior are very much a true representation of it. But this is not generally the case in biological and natural systems as well as in the engineering systems involving, for example, the components made of natural materials. We also observe that "noise" occurs randomly, i.e. we can not model them using the deterministic approaches. If we observe the system fairly accurately, and still we see randomness in spatial or temporal domains, then the "noise" is inherent and caused by system dynamics. In these instances, we refer to "noise" as randomness induced by the system.

There is a good example given by Øksendal et al. (1998) of an experiment where a liquid is injected into a porous body and the resulting scattered distribution of the liquid is not that one expects according to the deterministic diffusion model. It turns out that the permeability of the porous medium, a rock material in this case, varies within the material in an irregular manner. These kinds of situations are abound in natural and other systems, and stochastic calculus provides a logical and mathematical framework to model these situations. Stochastic processes have a rich repository of objects which can be used to express the randomness inherent in the system and the evolution of the system over time. The stochastic models purely driven by the historical data, such as Markov's chains, capture the system's temporal dynamics through the information contained in the data that were used to develop the models. Because we use the probability distributions to describe appropriate sets of data, these models can predict extreme events and generate various different scenarios that have the potential of being realized in the real system. In a very general sense, we can say that the probabilistic structure based on the data is the engine that drives the model of the system to evolve in time. The deterministic models based on differential calculus contain differential equations to describe the mechanisms based on which the model is driven to evolve over time. If the differential equations developed are based on the conservation laws, then the model can be used to understand the behavior of the system even under the situations where we do not have the data. On the other hand, the models based purely on the probabilistic frameworks can not reliably be extended to the regimes of behavior where the data are not available.

The attractiveness of the stochastic differential equations (SDE) and stochastic partial differential equations (SPDE) come from the fact that we can integrate the variability of the system along with the scientific knowledge pertaining to the system. In relation to the above-mentioned diffusion problem of the liquid within the rock material, the scientific knowledge is embodied in the formulation of the partial differential equation, and the variability of the permeability is modeled by using random processes making the solving of the problem with the appropriate boundary conditions is an exercise in stochastic dynamics. We use the term "stochastic dynamics" to refer to the temporal dynamics of random variables, which includes the body of knowledge consisting of stochastic processes, stochastic differential equations and the applications of such knowledge to real systems. Stochastic processes and differential equations are still a domain where mathematicians more than anybody else are comfortable in applying to natural and biological systems. One of the aims of this book is to explain some useful concepts in stochastic dynamics so that the scientists and engineers with a background in undergraduate differential calculus could appreciate the applicability and appropriateness of these recent developments in mathematics. We have attempted to explain the ideas in an intuitive manner wherever possible without compromising rigor.

We have used the solute transport problem in porous media saturated with water as a natural setting to discuss the approaches based on stochastic dynamics. The work is also motivated by the need to have more sophisticated mathematical and computational frameworks to model the variability one encounters in natural and industrial systems. The applications of stochastic calculus and differential equations in modeling natural systems are still in infancy; we do not have widely accepted mathematical and computational solutions to many partial differential equations which occur in these models. A lot of work remains to be done. Our intention is to develop ideas, models and computational solutions pertaining to a single problem: stochastic flow of contaminant transport in the saturated porous media such as that we find in underground aquifers. In attempting to solve this problem using stochastic concepts, we have experimented with different ideas, learnt new concepts and developed mathematical and computational frameworks in the process. We discuss some of these concepts, arguments and mathematical and computational constructs in an intuitive manner in this book.

1.2 Solute Transport in Porous Media

Flow in porous media has been a subject of active research for the last four to five decades. Wiest et al. (1969) reviewed the mathematical developments used to characterize the flow within porous media prior to 1969. He and his co-authors concentrated on natural formations, such as ground water flow through the soil or in underground aquifers.

Study of fluid and heat flow within porous media is also of significant importance in many other fields of science and engineering, such as drying of biological materials and biomedical studies. But in these situations we can study the micro-structure of the material and understand the transfer processes in relation to the micro-structure even though modeling such transfer processes could be mathematically difficult. Simplified mathematical models can be used to understand and predict the behavior of transport phenomena in such situations and in many cases direct monitoring of the system variables such as pressure, temperature and fluid flow may be feasible. So the problem of prediction can be simplified with the assistance of the detailed knowledge of the system and real-time data.

However, the nature of porous formation in underground aquifers is normally unknown and monitoring the flow is prohibitively expensive. This forces scientists and engineers to rely heavily on mathematical and statistical methods in conjunction with computer experiments of models to understand and predict, for example, the behavior of contaminants in aquifers. In this monograph, we confine our discussion to porous media saturated with fluid (water), which is the case in real aquifers. There are, in fact, two related problems that are of interest. The first is the flow of the fluid itself, and the second the transport of a solute introduced into the flow at a specific point in space.

The fluid flow problem is usually one of stationary flow, i.e, the fluid velocity does not change with time as long as external influences such as pressure remain constant. The overall flow rate (fluid mass per unit time) through a porous medium is well described by Darcy's law, which states that the flow rate is proportional to the pressure gradient. This is analogous to Ohm's law in the more familiar context of the flow of electric current. The coefficient of proportionality is a constant describing a property of the porous material, as is resistance for the case of an electrical conductor. The most obvious property of a porous material is that it partially occupies the volume that would otherwise be available to the fluid. This is quantified by defining the porosity ϕ of a particular porous medium, as the fraction of the overall volume that is occupied by the pores or voids, and hence filled by liquid for a saturated medium. Taking the porosity value separately, the coefficient in Darcy's equation is defined as the hydraulic conductivity of the medium.

The solute transport problem on the other hand, is a non-stationary problem: solute is introduced into the flow at a specific time and place, and the temporal development of its spatial distribution is followed. It is important in its own right, for example, to describe the propagation of a contaminant or nutrient introduced into an aquifer at some point. In addition, it can be used as an experimental tool to study the underlying flow of the carrier liquid, such as by observing the spread of a dye droplet, a technique also used to observe a freely flowing liquid. In free flow, the dye is carried along by the flow, but also gradually spreads due to diffusion on the molecular scale. This molecular scale or microdiffusion, takes place also in a static liquid because of the thermal motion of the fluid and dye molecules. It is well described mathematically by Fick's law, which postulates that the diffusive flow is proportional to the concentration gradient of the dye.

Past experience shows that when a tracer, which is a labeled portion of water which may be identified by its color, electrical conductivity or any other distinct feature, is introduced into a saturated flow in a porous medium, it gradually spreads into areas beyond the region it is expected to occupy according to micro diffusion combined with Darcy's law. As early as 1905 Slitcher studied the behavior of a tracer injected into a groundwater movement upstream of an observation well and observed that the tracer, in a uniform flow field, advanced gradually in a pear-like form which grew longer and wider with time. Even in a uniform flow field given by Darcy's law, an unexpectedly large distribution of tracer concentration showed the influence of the medium on the flow of the tracer. This result is remarkable, since the presence of the grains or pore walls that make up the medium might be expected to impede rather than enhance the distribution of tracer particles – as it does indeed happen when the carrier fluid is stationary. The enhanced distribution of tracer particles in the presence of fluid flow is termed hydrodynamic dispersion, and Bear (1969) described this phenomenon in detail.

Hydrodynamic dispersion is the macroscopic outcome of a large number of particles moving through the pores within the medium. If we consider the movement of a single tracer particle in a saturated porous medium under a constant piezometric head gradient in the x direction, we can understand the phenomenon clearly (Figure 1.1). In the absence of a porous medium, the particle will travel in the direction of the decreasing pressure (x- direction) without turbulence but with negligibly small Brownian transverse movements. (Average velocity is assumed low and hence, the flow field is laminar.) Once the tube in Figure 1.1 is randomly packed with, for example, solid spheres with uniform diameter, the tracer particle is forced to move within the void space, colliding with solid spheres and traveling within the velocity boundary layers of the spheres.

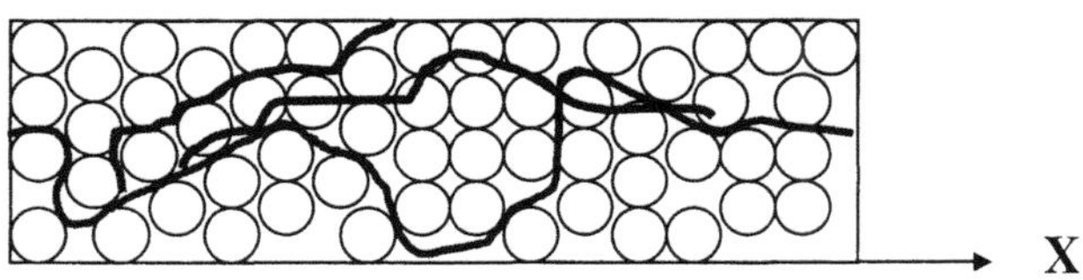

Figure 1.1 A possible traveling path of a tracer particle in a randomly packed bed of solid spheres.

As shown in Figure 1.1, a tracer particle travels in the general direction of x but exhibits local transverse movements, the magnitude and direction of which depend on a multitude of localized factors such as void volume, solid particle diameter and local fluid velocities. It can be expected that the time taken for a tracer particle to travel from one end of the bed to the other is greater than that taken if the solid particles are not present. If a conglomeration of tracer particles is introduced, one can expect to see longitudinal and transverse dispersion of concentration of particles with time. The hydrodynamic dispersion of a tracer in a natural porous formation occurs due to a number of factors. The variation of the geometry of the particle that constitute the porous formations play a major role in "splitting" a trace into finer "off-shoots", in addition, changes in concentration of a tracer due to chemical and physical processes, interactions between the liquid and the solid phases, external influences such as rainfall, and molecular diffusions due to tracer concentration. Diffusion may have significant effect on the hydrodynamic dispersion; however, we are only concerned with the effects of the geometry to larger extent and effects of diffusion to lesser extent. For the current purpose, in essence, the hydrodynamic dispersion is the continuous subdivision of tracer mass into finer 'offshoots', due to the microstructure of the medium, when carried by the liquid flowing within the medium. Because the velocities involved are low, one can expect molecular diffusion to have a significant impact on the concentration distribution of the tracer over a long period of time. If the effects of chemical reactions within the porous medium can be neglected, dispersion of tracer particles due to local random velocity fields, and molecular diffusion due to concentration gradients, are the primary mechanisms that drive the hydrodynamic dispersion.

1.3 Models of Hydrodynamic Dispersion

The basic laws of motion for a fluid are well known in principle, and are usually referred to as the Navier-Stokes equations. It turns out that the Navier-Stokes equations are a set of coupled partial differential equations that are difficult to solve even for flow in cavities with relatively simple geometric boundaries. It is clearly impossible to solve them for the multitude of complex geometries that will occur in a detailed description of the pore structure of a realistic porous medium. This level of detail is also not of practical use; what is desired is a description at a level of detail somewhere intermediate between that of Darcy's law and the pore level flow. Different approaches to achieve this have been described in literature (e.g. Taylor, 1953; Daniel, 1952; Bear and Todd, 1960; Chandrasekhar, 1943). These approaches can broadly be classified into two categories: deterministic and statistical.

In the deterministic models the porous medium is modeled as a single capillary tube (Taylor, 1953), a bundle of capillary tubes (Daniel, 1952), and an array of cells and associated connecting channels (Bear and Todd, 1960). These models were mainly used to explain and quantify the longitudinal dispersion in terms of travel time of particles and were confined to simple analytical solutions (Bear, 1969). They have been applied to explain the data from laboratory scale soil column experiments.

Statistical models, on the other hand, use statistical theory extensively to derive ensemble averages and variances of spatial dispersion and travel time of tracer particles. It is important to note that these models invoke an ergodic hypothesis of interchanging time averages with ensemble averages after sufficiently long time, and the law of large numbers. By the law of large numbers, after a sufficiently long time, the time averaged parameters such as velocity and displacement of a single tracer particle may replace the averages taken over the assembly of many particles moving under the same flow conditions. Bear (1969) questioned the validity of this assumption arguing that it was impossible for a tracer particle to reach any point in the flow domain without taking the molecular diffusion into account.

In statistical models, the problem of a cloud of tracer particles traveling in a porous medium is reduced to a problem of a typical single particle moving within an ensemble of randomly packed solids. Characteristic features of these models are: (a) assumed probability distributions for the properties of the ensemble; (b) assumptions on the micro dynamics of the flow, such as the relationships between the forces, the liquid properties and velocities during each small time step; (c) laminar flow; and (d) assumed probability distributions for events during small time step within the chosen ensemble. The last assumption usually requires correlation functions between velocities at different points or different times, or joint probability distributions of the local velocity components of the particle as functions of time and space, or a probability of an elementary particle displacement (Bear, 1969).

Another modeling approach that has been used widely is to consider the given porous medium as a continuum and apply mass and momentum balance over a Representative Elementary Volume (REV) (Bear et al., 1992). Once the assumption is made that the properties of the porous medium, such as porosity can be represented by average values over the REV, then the mass and momentum balances can be applied to a REV to derive the governing partial differential equations which describe the flow in the medium. Since the concept of the REV is central to this development, it is important to summarize a working model based on this approach.

1.4 Modeling Macroscopic Behavior

1.4.1 Representative Elementary Volume

The introduction of a REV is once more analogous to the approach followed in electromagnetic theory, where the complexities of the microscopic description of electromagnetic fields at a molecular level, is reduced to that of smoothly varying fields in an averaged macroscopic continuum description. The basic idea is to choose a representative volume that is microscopically large, but macroscopically small. By microscopically large, we mean that the volume is large enough that fluctuations of properties due to individual pores are averaged out. Macroscopically small means that the volume is small enough that laboratory scale variations in the properties of the medium is faithfully represented by taking the average over the REV as the value associated with a point at the center of the REV. For this approach to be successful, the micro- and macro-scales must be well enough separated to

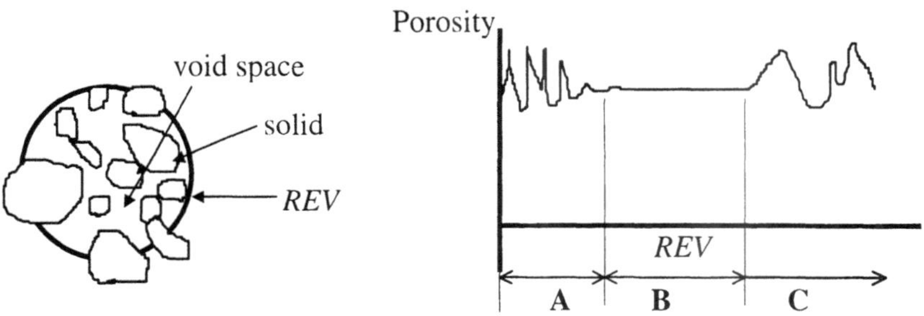

Figure 1.2 Variation of porosity with Representative Elementary Volume (REV).

allow an intermediate scale – that of the REV – at which the exact size and shape of the REV makes no difference.

Porosity is defined as the ratio between the void volume and the overall volume occupied by the solid particles within the REV. The variation of porosity with the size of REV is illustrated in Figure 1.2 (Bear et al., 1992). The fluctuation in porosity values in region A shows that the REV is not sufficiently large to neglect the microscopic variations in porosity. If the porous medium is homogeneous, porosity is invariant once region B is reached, which can be considered as the operational region of REV for mass and momentum balance equations. For a heterogeneous porous medium, porosity variations still exist at a larger scale and are independent of the size of REV (Region C).

Once the size of REV in the region B is established for a given porous medium, macroscopic models can be developed for the transport of a tracer (solute). The variables, such as velocity and concentration, are considered to consist of a volume-averaged part and small perturbations, and these small perturbations play a significant role in model formulations (Gray, 1975; Gray et al., 1993; Hassanizadeh and Gray, 1979; Whitaker, 1967).

1.4.3 Review of a Continuum Transport Model

To make the discussion of the transport problem more concrete, we turn our attention to an example with a simple geometry. Consider a cylindrical column of internal radius R with the Cartesian coordinate system as shown in Figure 1.3. The column is filled with a solid granular material and it is assumed that the typical grain diameter (l_d)<< R. Assuming that the porous matrix is saturated with a liquid of density, ρ, the local flow velocity of the liquid with respect to the stationary porous structure and the local concentration of a neutral solute in the fluid are denoted by $v(x,y,z,t)$ and $c(x,y,z,t)$, respectively. The REV or averaging volume (δV) for this system is a cross sectional volume of the column of some width, $\Delta\varsigma$. It is assumed that δV is sufficiently large so that statistical averages are insensitive to small variations in δV (Rashidi et al., 1996).

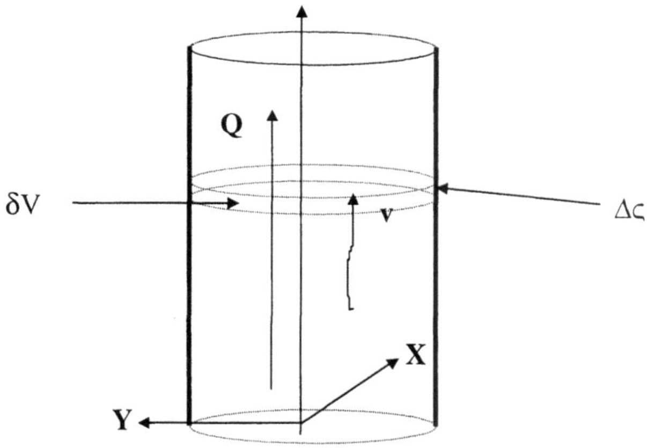

Figure 1.3 Geometry for the cylindrical column flow model.

The volume average of a pore scale quantity, ψ, associated with the liquid phase is defined by (The notation for the model is adopted from Rashidi et al. (1996).)

$$\langle \psi(z,t) \rangle = \frac{1}{\pi R^2 \Delta\zeta} \int_{-\Delta\zeta/2}^{\Delta\zeta/2} \int_A \psi(x,y,z+\zeta,t) \ \gamma \ (x,y,z+\zeta) \ dA \ d\zeta \qquad (1.1)$$

where A represents the cross-sectional area of the column and γ is an indicator function which equals 1 if the point (x,y,z+ς) lies in the void space, and zero otherwise. The cross sectional porosity, φ(z), is obtained by setting ψ=1 in equation (1.1).

The volumetric flux, q_z is defined as the volume of fluid passing a point z per unit time. Since the microscopic momentum flux (momentum per unit area) carried by the fluid at any point (x,y,z) is given by $\rho\underline{v}$ the macroscopic momentum flux for an incompressible fluid is given by

$$\langle \rho\underline{v} \rangle = \rho\underline{q} = \rho q_z \underline{k} \qquad (1.2)$$

where $\underline{k}$ is a unit vector along the z- axis. The total volumetric flux through the cross section is given by

$$Q = q_z(\pi R^2), \tag{1.3}$$

and the mean velocity can be defined by,

$$\bar{v} = q_z \underline{k} / \varphi \ . \tag{1.4}$$

The instantaneous, local solute flux consists of a contribution (cv) representing solute carried along by the liquid flow (advection), and a diffusion contribution proportional to the concentration gradient. Because the solute flux is non-stationary, conservation of solute mass is expressed by the time-dependent equation of continuity.

Using averaging theorems, this can be reduced to the following one-dimensional macroscopic mass balance equation for the solute (Thompson et al., 1986):

$$\frac{\partial(\varphi\bar{c})}{\partial t} + \frac{\partial\overbrace{(\varphi\bar{v}_z\bar{c})}^{A}}{\partial z} + \frac{\partial\overbrace{(\varphi J_z)}^{B}}{\partial z} - \frac{\partial}{\partial z}\overbrace{\left(\varphi D_m(\frac{\partial\bar{c}}{\partial z} + \tau_z)\right)}^{C} = 0 \ . \tag{1.5}$$

The various terms in this equation can be interpreted as a rate of change of the intrinsic volume average concentration, balanced by the spatial gradients of the terms, A, B, and C respectively.

Term A represents the average volumetric flux of the solute transported by the average flow of fluid in the z-direction at a given point in the porous matrix, (x,y,z). But the total solute flux at a given point is the sum of the average flux and the fluctuating component due to the velocity fluctuation above the mean velocity, $\overline{v_z}$. We introduce the perturbation terms of velocity (v_z') and concentration (c'), each of which represents the difference between the microscopic quantity evaluated at (x, y, z+ς) (within a REV), and the corresponding intrinsic average evaluated at z. In terms of these, the fluctuating component of the flux is given by:

$$J_z(z,t) = \overline{v_z'}\,\overline{c'} \ . \tag{1.6}$$

The terms A and B are called the mean advective flux and the mean dispersive flux, respectively. Making the following assumption for the dispersive flux, based on plausibility arguments, often circumvents the need for the detailed knowledge of the fluctuation terms:

$$J_z(z,t) \approx -\alpha_L \overline{v}_z \frac{\partial \overline{c}}{\partial z}. \tag{1.7}$$

Why would the dispersive flux be proportional to a concentration gradient? The velocity fluctuations that underlie this term, must by definition sum to zero over an REV. They therefore have the effect of exchanging fluid elements from different spatial locations rather than a net flow of liquid. Nevertheless, if the concentration is different at these locations, this will have the effect of a net transport of solute mass, while if the concentration gradient is zero, no net solute mass will be transported. Any existing concentration gradient will tend to be decreased by such a mixing mechanism, which accounts for the negative sign in equation (1.7). The proportionality to the average velocity assumed in that equation is plausible since the fluctuations under discussion do not have an independent origin but arise from deflection of fluid elements away from the averaged flow by the pore walls. Therefore, these fluctuations will be zero in the absence of flow and progressively increase as the flow increases. Both of these dependencies might conceivably be more complex, but the assumption of simple proportionality in equation (1.7) seems a reasonable working assumption.

The proportionality constant in the equation, α_L, is defined as the dispersivity of the porous medium. Dimensional considerations show that it has the dimensions of a length, and an obvious hypothesis is to connect it with the granularity scale of the medium, i.e. with a typical particle diameter. Although of a different physical origin, the form of equation (1.7) is reminiscent of Fick's law and the similarity is often emphasized by defining the product $D = \alpha_L \overline{v}_z$ as the dispersion coefficient in analogy to the diffusion coefficient, D_m.

The term C in equation (1.5) is the modified Fick's law for micro diffusion of the solute. A modification is needed, because even in the absence of any averaged flow of the liquid, molecular diffusion will be affected by the presence of the porous medium. In particular, an increased path length is needed by a diffusing particle, to reach the same displacement relative to its starting position as in free flow conditions. This increased path length is a measure of the degree of tortuous paths present within the porous medium. The term τ_z that arises from the mathematical averaging is in fact the total surface area perpendicular to the flow presented by the grains of the porous medium within the REV, and weighted by the concentration at each point. It can be seen as a measure of the total barrier to diffusion caused by the tortuosity of the medium.

While the tortuosity is essentially a geometric concept, its effect appears in equation (1.5) as a modification of the diffusion term and hence τ_z is referred to as the diffusive tortuosity. It is reduced to a dimensionless parameter β that characterizes the structure of the medium by writing

$$\tau_z(z,t) \approx -\beta \frac{\partial \bar{c}}{\partial z}. \tag{1.8}$$

The assumption that this is also proportional to the concentration gradient, may be justified by considering that the existence of a barrier only has an effect if there is a non-zero concentration gradient at a particular position x. Substituting equations (1.7) and (1.8) into (1.5), yields

$$\frac{\partial}{\partial t}\left(\varphi \bar{c}\right) + \frac{\partial}{\partial z}\left(\varphi \bar{v}_z \bar{c}\right) - \frac{\partial}{\partial z}\left[\left[\varphi D + \varphi D_m (1-\beta)\right] \frac{\partial \bar{c}}{\partial z}\right] = 0. \tag{1.9}$$

From this expression it becomes clear that the entire range of plausible effects on diffusion by the tortuosity of the medium, ranging from negligible effect when the medium allows fluid elements to proceed along free flow fluid lines, to complete blocking of diffusion when fluid elements are forced by the pore geometry into infinite path lengths to cover a finite displacement, can be represented by letting β range over the values 0 to 1 respectively.

The sum $D_H = D + D_m(1-\beta)$ is called the coefficient of hydrodynamic dispersion. In many cases, $D \gg D_m$ such that $D_H \approx D$ (Rashidi et al., 1996). Equation (1.9) is the commonly-used working model for solute transport in porous media and is usually called the advection-dispersion equation. Note that it is dependent on the Fickian assumptions expressed by equations (1.7) and (1.8).

A mental picture of the interplay between dispersion and micro diffusion can perhaps be facilitated by the following analog. Consider a spoonful of sugar at the bottom of a teacup. As the sugar dissolves, it will spread slowly through the cup as a result of micro diffusion. If the tea is stirred, this action introduces advection, moving sugar to other parts of the liquid; and in so doing, larger gradients in the concentration are produced and diffusion is enhanced, leading to a very quick spreading of the sugar. In a porous medium, the deflection of the flow by pore walls plays the role of the stirring, but unlike the case of the teacup the pore walls also slows down the diffusion. The net effect is nevertheless an enhancement of the spreading beyond that produced by diffusion alone.

We have seen that the division of solute flux into advective and dispersive components is a result of the mathematical division of velocity and concentration into the mean and perturbation terms. Then the perturbations are linearized through Fickian assumptions making the model deterministic so that we can understand the behavior of solute dispersion. However, this liberalization changes the inherent random character of the problem into a deterministic one. In addition we observe that:

- the coefficient of hydrodynamic dispersion depends upon the velocity fluctuations induced by the pore structure; therefore, it can be expected to be scale dependent;
- Fickian type assumptions are made for dispersive flux as well as for diffusive tortuosity. These assumptions are based on plausibility arguments, not justified on the basis of detailed microscopic theory as in the case of micro-diffusion;
- there are only two physical phenomena involved in the solute transport in a porous medium: the solute is carried by the flowing fluid (advection) and, if the velocities are very small, micro-diffusion can occur as described by the Fick's law; and,
- the working model described above can only be applied to homogeneous porous media where a representative elementary volume can be defined.

1.5 Measurements of Dispersivity

Rashidi et al. (1996) experimentally investigated the solute transport in a bed of packed polymethylmethacryle (PMMA) plastic spherical beads of 0.31 cm diameter contained in a cylindrical column having 4.5 cm diameter. Their experiments are designed to test the validity the Fickian assumptions as well as several other measurement and estimation techniques commonly used in applying the advection-dispersion equation, by performing direct measurement of the pore-scale velocity and concentration fields.

In their experiment, the beads and the fluid were chosen to be transparent and have the same refractive index to allow direct optical probing at any point within the porous system. Fluid velocity was measured by introducing fluorescent micro-spheres of 6.5 μm diameter into the flow, and measuring their velocity by observing their movement across a series of illumination planes. Using laser optics, the illumination planes were prepared to a resolution orders of magnitude smaller than the grain and pore diameters, allowing non-invasive velocimetry at essentially point-like resolution within the pores. In a similar way, concentrations were measured non-invasively at the same point grid by introducing a fluorescent dye as the solute tracer. Their results are reproduced in Table 1.1.

From their measured velocities, the fluctuations and hence the dispersive flux as defined in equation (1.6) could be directly calculated by averaging. From the observed concentrations, the gradient of the average was similarly found, and the ratio of these is a direct measurement of the dispersion coefficient and hence the dispersivity.

Table 1.1 Summary of measured or estimated dispersivities and dispersion coefficients (Rashidi et al., 1996).

Method	Dispersivity α_L (cm)	Dispersion Coefficient D (cm^2/min)	Assumptions and comments
Direct experiment	0.12	0.018	Rashidi et al. (1996); Constant velocity experiments to get α_L; D found directly.
Length-scale arguments			
$\alpha_L \approx 0.5\ l_d$	0.15	0.023	$Pe = \overline{v}_z l_d / D_m \approx 75$
$\alpha_L \approx 1.8\ l_d$	0.54	0.081	
$\alpha_L \approx l_d$	0.30	0.045	Bear (1972) pp 609
Breakthrough slope methods	0.20	0.031	Constant velocity; constant porosity; square Initial Conditions (IC); averages over 28 locations
Numerical curve fits:			Constant velocity
Square IC	0.20	0.030	Constant porosity; square IC
Slope IC	0.12	0.018	Constant velocity; porosity

As mentioned in the previous section, the dispersivity is essentially a length scale and as the primary length scale that characterizes the structure of a porous medium is the diameter of its pores or the particles that makes it up, it is reasonable to expect a relation between these. This line of argument has been elaborated by Bear (1972). He proposed a direct proportionality of the dispersivity to the grain diameter, but with different values for the proportionality constant according to the value of the Peclet number. The Peclet number is a dimensionless ratio that expresses the relative importance of advection, and diffusion or dispersion, for particular flow conditions. It is given by $Pe = \overline{v}_z l_d / D_m \approx 75$ for the experiment. This value is intermediate between zones distinguished by Bear, and Rashidi et al. compares the values yielded for the dispersivity by both zones as well as an average.

In the breakthrough slope methods, solute concentration is measured as a function of time at a fixed position when a solute front advances past the point. Comparing the measured curve with analytic solutions of the dispersion-advection equation for an idealized step function initial concentration, a value for the dispersion coefficient is extracted. This is less detailed than the complete measurement of concentration and velocity fields in the direct method, but is similar to field measurements in macroscopic and aquifer experiments.

Finally, D is similarly estimated by comparison with numerically calculated solutions of the dispersion-advection equation (the curve-fitting solutions). Two cases are considered: a square step initial condition, and one where the initial concentration already contains some dispersion by assuming a sloped step. The latter provides for insufficient experimental control to produce the idealized mathematical initial condition. Values of dispersivity and dispersion coefficient obtained from different approaches for the same porous medium vary considerably as shown in Table 1.1.

Given the advanced laser beam measurement techniques employed in the method (Rashidi et al., 1996) it is reasonable to assume that the values reported by the direct method are 'true' values. Then the length scale arguments tend to overestimate the dispersivity, although giving reasonable agreement at the lower end of the range of estimates. A plausible explanation would be that the proportionality coefficients proposed by Bear, were found for natural media like sand containing irregularly shaped grains, and would be expected to overestimate dispersion caused by the smooth spheres in the experiment. Hence length scale arguments are good for first estimates but may need further refinement by measuring actual dispersion.

Applying the standard methods based on breakthrough curves, both analytical and numeric, also tended to overestimate the dispersion. In part, this may be due to the assumption of a physically unrealistic initial condition, as good agreement could be obtained in the last row in Table 1.1 by fitting a slope for the assumed initial condition step. However, Rashidi et al. also calculated the dispersive flux over the whole of the measurement volume, as a function of time, from the numerical solution. When this is compared to the true dispersive flux calculated by applying equation (1.6) to the measured velocities, the dispersive flux calculated from the advection-dispersion equation overestimates the true value by a factor 3 for D=0.03 cm^2/min as obtained from the breakthrough curve analysis. Both reducing D to the true value of 0.018 cm^2/min, and introducing a sloped step initial condition, reduced the discrepancy to a factor 2. Rashidi et al. reported additional attempts to further reduce the discrepancy by allowing for spatial variability in the porosity or dispersivity, but without success. This strongly suggests that the Fickian assumption for the dispersive flux, and/or for the tortuosity, is not fully satisfied in the experiment.

Another point demonstrated by Table 1.1, is the dominance of dispersion over diffusion in the experiment. Typical diffusion coefficients in aqueous solutions vary from 0.00026 cm^2/min for a small non-ionic molecule such as sucrose, to 0.002 cm^2/min for a strong electrolyte like HCl (see Weast (1972)). The much higher dispersion coefficients in Table 1.1 shows that dispersion far outweighs micro diffusion in this experiment. A high dispersion coefficient is a result of relatively high velocity perturbations, i.e. increased randomness in the velocity field. We conclude that the random component in the velocity field plays a significant role in determining the concentration field, even in a homogeneous porous medium tested under laboratory conditions.

Moroni and Cushman (2001) conducted three-dimensional particle tracking velocimetry (3D-PTV) experiments for a homogenous porous medium of Pyrex 1.9-cm spheres and glycerol. Air bubbles diffused into the system was used as tracer particles and they were drained along with glycerol. Three-dimensional trajectories of the tracer particles were constructed and analyzed based on the data from the experiments. They found that the longitudinal velocity distributions were neither Gaussian nor lognormal, but the transverse velocity distributions were Gaussian. Dispersion coefficients in the transverse directions approaches to zero with increasing time but some trajectories showed sharp increases in dispersion coefficient at the beginning and others did not show significant increases. Normalized longitudinal velocity covariances for different flow rates had similar exponentially decaying behavior with respect to time. Longitudinal dispersion coefficient tends to be constant at a given flow rate after a relatively small time indicating that the flow in longitudinal direction can be considered Fickian. However, this can not be said about the other two directions which makes the application of the Fickian assumptions as a general rule even in a homogeneous medium questionable.

1.6 Flow in Aquifers

1.6.1 Transport in Heterogeneous Natural Formations

Field experiments show that spatial heterogeneity is the most significant factor affecting dispersion of solute in natural formations such as aquifers (Anderson, 1979; Gelhar et al., 1985; Freyberg, 1986). Dagan (1988) referred to the experimental results depicting large scale spatial distribution of hydraulic conductivity at the Borden site. Large scale irregular variations in hydraulic properties, such as hydraulic conductivity and porosity, lead to a high degree of uncertainty of parameters in transport equations based on the continuum approach, whose application to solute transport in natural formations is questionable (Dagan, 1988). Dagan (1988) concluded that the concentration of a solute can be considered as a random variable, which can be described by its statistical moments; the ergodic hypothesis can not be applied except in a limited sense in describing some gross features such as the motion of the center of gravity and the second order spatial moment of the solute body; it is extremely difficult to predict accurately the point value of concentration in an irregularly distributed body of solute within the natural formations; the expected value of concentration does not necessarily satisfy a advection-dispersion type equation, such as equation 1.9, based on the continuum approach; and even if the latter is satisfied, the dispersion coefficient increases with the travel time reaching an asymptotic value. In other words, the flow is non-Fickian[1] during the preasymptotic period.

Dagan (1988, 1990) developed a mathematical formulation to predict the spatial moments of concentration plumes in aquifers introducing perturbation terms to describe random variations of variables such as concentration and velocity. It is usually assumed that a solute is transported by advection described by the Darcian steady velocity and by pore scale diffusion. The governing equation is the 3-dimensional generalization of equation (1.9) and is given by,

$$\frac{\partial C}{\partial t}+V_j\frac{\partial C}{\partial x_j}=\frac{\partial}{\partial x_j}\left(D_{d,jl}\frac{\partial C}{\partial x_l}\right)\quad j,l=1,\ 2,\ 3 \tag{1.10}$$

---

[1] When the flow can be described by a advection-dispersion type equation, the transport is said to be Fickian.

Here $C=\varphi\,\bar{c}$ is the concentration per unit volume of aquifer, $x = (x_1,x_2,x_3)$ is the coordinate vector, and V is the Eulerian velocity vector according to Darcy's law:

$$V = -\frac{K}{\varphi}\nabla\phi \ . \tag{1.11}$$

In Darcy's equation, K is the hydraulic conductivity; φ is the effective porosity and ϕ is the piezometric head. Since Darcy's equation gives V as a constant, it was factored out of the differentiation in the second term of equation (1.10). In order to provide for anisotropy in the dispersion, $D_{d,jl}$ in that equation is taken as a tensor rather than a scalar, and represents the effect of pore scale dispersion (Dagan, 1988). The standard summation convention for repeated indices was used.

Dispersive flux due to pore scale velocity variations is lumped into the right hand term in equation (1.10) through a diffusion tensor and the mean velocity given by Darcy's law is used in the advection term. Dagan (1988) assumed constant components for the pore scale dispersive tensor, $D_{d,jl}$, and introduced macro scale variations as perturbation terms to the velocity and concentration. After a non-trivial mathematical analysis he arrived at the expected values of the spatial second moments of a solute plume within a stationary heterogeneous porous structure.

Cushman (1987) stated that the transport process in natural formations can not be modeled by the advective-dispersive equation because of stochastic (random) fluctuations in flow velocity due to natural heterogeneity in the pore structure and failure of Fick's type diffusion equation to describe the pore scale dispersion.

A particular problem in the application of the advection-dispersion equation and attributed by many authors to the natural heterogeneity is the so-called scale dependence of the dispersivity. For example, Pickens and Grisak (1981) found from laboratory measurements of a 30-cm column of a certain aquifer, the longitudinal dispersivity to be α =0.035 cm. In single well field tests on the same aquifer, a value of α =3 cm was obtained for a flow length of 3.1 m and α =9 cm when the solute plume traveled 5 m. In two-well tests with wells located 8 m apart, the value α =50 cm was obtained.

Such variability of a "constant" is very disturbing in a model of which a major merit is intended to be the reduction of a complex phenomenon to simple materials constants. At best, it indicates the presence of a key variable that was not recognized in the first formulation of the model. If this variable can be identified, the situation might be rescued. One attempt to do this, was to express α as proportional to the flow length. The results of Lallemand-Barres et al. (1978) fits such a relation to the measured values for a number of natural aquifer materials; despite considerable scatter, a reasonable fit over a scale range of 10^3 was obtained. However, when the range was extended to about 10^5 by Gelhar (1986), no simple relationship appears tenable any more.

Many authors concur that the root of the problem is that natural aquifers are not homogeneous. The measured values of hydraulic conductivity for different geological materials are found to vary by up to 9 orders of magnitude; porosity differs by more than 2 orders of magnitude. One can hardly expect the value of such a sensitive quantity to remain constant over the extent of a natural aquifer even if composed of a single mineral. This is borne out by actual measurements reported by Gelhar (1986) and by Sudicky (1986) showing spatial variations over one order of magnitude.

Scale dependence of the dispersion process has been examined in the light of spatial distribution of hydraulic conductivity in many studies (Serrano, 1988). However, Serrano (1988) has cited other researchers' work, which mentioned the inconsistencies of the approaches to solve the transport equation in their review article. They have concluded that the advective-dispersive model neither explained the scale effect of the dispersion process nor elucidated the transient behavior of solute concentration. In addition, Serrano (1988) summarized the main difficulties with the existing solutions. The assumption that the perturbations are small random variations in the perturbation expansion solutions (Gelhar and Axness, 1983; Dagan, 1984) is not valid in a differential equation unless the variables have small variance. One could argue that if the variances of the processes involved are indeed small as required by the perturbation solutions, then a completely deterministic model could be used as a sufficient tool for prediction purposes. Cushman (1987) made similar conclusions after an in-depth analysis of the perturbation solutions provided by Gelhar and Axness (1983).

The assumption that hydraulic conductivity is the only source of uncertainty in many existing solutions has to be challenged. Dispersion in natural formations can not only be explained by the variability of hydraulic conductivity; and if this is feasible the large-scale hydraulic conductivity can safely be described by well defined deterministic functions. Some of the solutions are only valid in the steady state of the concentration field.

The assumed ergodicity in the hydraulic conductivity field in some solutions does not necessarily imply that the concentration field is ergodic. The temporal and spatial variations of the parameters affecting the aquifer heterogeneity significantly make the analytical solutions of dispersion equations almost impossible. However, recent attempts to solve the advective-dispersion equation for heterogeneous aquifers are noteworthy (Serrano, 1996). Serrano (1996) solved the three-dimensional solute continuity equation in a large domain neglecting the micro diffusion and using a decomposition method. However, the probability law of the velocity fields needs to be provided to solve the solute transport equation for a particular aquifer.

1.7 Computational Modeling of Transport in Porous Media

As we have seen in the previous discussion on the solute transport in heterogeneous media, the problem of predicting the solute concentration of a plume is complicated by many factors. Usually the nature of the porous formation is not known in detail but in some aquifers we can know the approximate composition of the medium, its particle sizes and their distribution. But in many cases, values of parameters such as hydraulic conductivity and porosity can only be measured at a limited number of points in the field. Spatial variability of hydraulic conductivity, for example, can not be known experimentally to our satisfaction.

If our mathematical models incorporating the possible variabilities cannot be solved without significant simplifications and assumptions, and the experiments on real aquifers is prohibitively expensive, where do we turn to understand more about the phenomenon? Computer models built on basic conservation equations incorporating variabilities modeled by known processes would provide us with a tool for experimentation. The whole purpose of the exercise is to develop a tool by looking at the fundamentals of the solute transport process and develop them into mathematical equations and logical relationships with a high degree of flexibility, to incorporate the stochasticity in parameters such as hydraulic conductivity and porosity. Once we have such a tool, we can conduct computational experiments and investigate how the solute behaves within a porous medium. As we have seen before, velocity of a solute particle in a porous body is a random variable and the variability is very much influenced by the pore structure of the body.

Equation (1.7) models the fluctuating component of the solute flux in a representative elementary volume using a Fickian-type spatial gradient of the mean concentration. The dispersion coefficient is defined as the coefficient relating the gradient of the mean concentration to the fluctuating component of the flux. The dispersion coefficient itself is expressed as a multiplication of the diffusivity and the mean velocity, and the diffusivity is found to be scale dependent. This could be expected. The Fickian relationship itself is an attempt to linearize a random process with respect to quantities such as particle sizes. As shown by Rashidi et al. (1996), even though this works fairly well to predict concentration, discrepancies in the dispersive flux arise even when the porous bed is very homogeneous. So when particle sizes are not uniform, the fluctuating component of the flux may become highly random and modeling that component using a Fickian model is not desirable. We are left with the option of modeling velocity as a random variable, and the degree of randomness depends on the pore structure.

The above discussion leads us to model the transport phenomenon in natural formations by treating the variables and the parameters involved in the processes as stochastic variables each having its own probability law, and when they satisfy the conservation laws, subsequent differential equations have to be solved by using stochastic differential calculus and appropriate numerical methods.

It is appropriate here to point out the difference between a deterministic differential equation solved for random boundary or initial conditions, and a stochastic differential equation. If for example in a time domain differential equation, the initial conditions are inaccurately known, it is in principle possible to solve the differential equation for a large number of different initial conditions, and use the probability distribution of such initial conditions to find the probability distribution of the solutions. Equivalently, one may specify the initial condition as an average value, modified by a random perturbation, and find the effect of the perturbation by operating on it with the deterministic differential operator. The statistically based studies of aquifer flow discussed above are mainly of this type. However, it is quite a different matter when the differential operator is itself stochastic, e.g. containing coefficients that undergo random variations during the interval between the initial condition and the time for which the solution is found. This much more difficult case is the domain of stochastic differential equations.

The problem of flow in a heterogeneous porous medium, is of the latter type. For example, in Darcy's equation, if there are random variations in the hydraulic conductivity over the spatial extent of the aquifer, a fluid element encounters continually changing values of the driving coefficient that determines its displacement as it follows a path through the porous medium. This is true of macro scale variations in the hydraulic conductivity, or the porosity, or both.

Moreover, this conceptual framework is just as applicable at the micro scale – pore level – length scale. If we consider the *gedanken-experiment* of observing the movement of a fluid element in a transparent porous medium, it will describe a path similar to Brownian motion. Brownian motion, indeed, represents the simplest possible stochastic differential equation.

This raises one of the main ideas to be explored in this book: that just as "real" Brownian motion is intimately connected with molecular scale diffusion, pore-scale Brownian motion may be used as a model of dispersion in a porous medium. In this way we will show that Fickian assumptions may be avoided and a more fundamental description addressing, for example, the scale dependence problem can be attempted. We are incorporating Brownian motion, a mathematical object or process representing variability in the system and the conservation laws in the stochastic dynamical description of the problem.

Before this idea can be implemented, it is first necessary to outline aspects of the theory of stochastic differential equations and some essential elements of stochastic processes that will be used. A comprehensive review of stochastic differential equations and related mathematical development are given by Kloeden and Platen (1992), Øksendal (1998) and their definitions and notation are extensively used. Reader may also refer to Klebaner (1998) to learn stochastic calculus with applications in finance.

Chapter 2

A Brief Review of Mathematical Background

2.1 Introduction

In the previous chapter, we saw that the deterministic formulation of the hydrodynamic dispersion requires us to make linearizing assumptions, which give rise to the coefficients that are dependent on the scale of the problem. This also makes us think of ways by which we could model the randomness introduced by the system, geometry of porous media, in the case of the hydrodynamic dispersion, without having to resort to any linearizing assumptions. Recent advances in stochastic calculus provide us with useful concepts and tools to model systems in much more natural way. We will attempt to review the pertinent concepts and methods in stochastic calculus prior to discussing stochastic differential equations later in Chapter 4 in some detail. However, to motivate the discussion on stochastic calculus we will first present the behavior of a simple model based on stochastic calculus. These type of examples can be found in Kloden and Platen (1992) and Øksendal (1998). We will outline the differences between the solutions of stochastic differential equations and ordinary differential equations using an example from population dynamics. In the next chapter we will discuss the models by using computer simulations to show the complexity of the solutions even though the solutions themselves appear to be straightforward mathematical expressions.

Let us consider a population growth model where the assumption is made that the rate of change of a population ($X(t)$) is linearly related to the population itself. We can write the equation as,

$$\frac{dX(t)}{dt} = \alpha(t)X(t), \tag{2.1}$$

where $\alpha(t)$ is a time dependent coefficient.

If $\alpha(t)$ is a constant ($\alpha(t)= \alpha_0$) then equation (2.1) has the solution,

$$X(t) = X_0 e^{\alpha_0 t}, \tag{2.2}$$

where X_0 is the initial population (i.e. the population at $t = 0$).

If $\alpha(t)$ is affected by unknown processes of the system to make it "noisy", then it can be expressed as the sum of an average rate constant $r(t)$ and a noise term with an amplitude $\sigma(t)$:

$$\alpha(t) = r(t) + \sigma(t)\, W_t\ . \tag{2.3}$$

The "noise" term W_t will be different at different times, but is not uniquely determined by the t value. That is why, for the time being, we use a subscript with W rather than a function notation, and will expand on that below. We generalize the discussion by considering the case where the rate at which the population changes, depends on the population in a more complicated way than a mere proportionality; arriving at the equation

$$\frac{dX_t}{dt} = b(t, X_t) + \sigma(t, X_t) W_t\ . \tag{2.4}$$

Equation (2.1) together with equation (2.3) represents the special case of equation (2.4) when $b(t,X_t) = r(t)\, X_t$ and $\sigma(t,X_t) = \sigma(t)\, X_t$. In equation (2.4) we have changed from $X(t)$ to X_t ; this is to signify that once $\alpha(t)$ is "noisy", it influences X to be "noisy" as well. In other words, X will no longer have a unique value at a given time and is not a simple function any more.

It is appropriate at this time to introduce the concepts of stochastic variables and processes. An ordinary differential equation, such as equation (2.1), defines a functional relationship between X and t that allows us to calculate a <u>deterministic variable</u> X uniquely for a given time, by a formula such as equation (2.2).

A <u>stochastic variable</u> Y, on the other hand, is one that does not have a unique value; it can have any one out of a collection of values. We assign a unique label ω to each possible value of the stochastic variable, and set Ω to denote the collection of all such values. In some cases, such as when Y represents the outcome of throwing dice, Ω may be a finite set of discrete numbers. In others, such as when Y is the instantaneous position of a particle undergoing

Brownian motion, it may be a continuous range of real numbers. If a particular value y_ω is observed for Y, this is called an event F. In fact, this is only the simplest prototype of an event; other possibilities might be that the value of Y is observed not to be y_ω (the complementary event), or that a value within a certain range of ω values is observed. The set of all possible events is denoted by $\mathcal{F}$. The final ingredient needed for Y to qualify as a stochastic variable, is that even though the outcome of a particular observation of Y is unpredictable, the probability of observing y_ω must be determined by a probability function $P(\omega)$. By using the standard methods of probability calculus, this implies that a probability $P(F)$ can also be assigned to compound events F e.g. by appropriate summation or integration over ω values. For this to work, $\mathcal{F}$ must satisfy the criteria that for any event F in $\mathcal{F}$, its complement F^c must also belong to $\mathcal{F}$, and that for any subset of F's the union of these must also belong to $\mathcal{F}$. The explanation above of what it means to call Y a stochastic variable, is encapsulated in formal mathematical language by saying "Y is defined on a probability space $(\omega, \mathcal{F}, P)$".

In describing physical systems, deterministic variables usually depend on additional parameters such as time. Similarly, a stochastic variable may depend on an additional parameter t (for example, the probability may change with time, i.e. $P(y_\omega, t)$). The collection of stochastic variables, Y_t , is termed a stochastic process. The word 'process' suggests temporal development and is particularly appropriate when the parameter t has the meaning of time, but mathematically it is equally well used for any other parameter, usually assumed to be a real number in the interval $[0,\infty]$. The label ω is often explicitly included in writing the notation $Y_t(\omega)$, for an individual value obtained from the set of Y-values at a fixed t. Conversely, we might keep ω fixed, and let t vary; a natural notation would be to write $Y_\omega(t)$. In physical terms, one may think of this as the set of values obtained from a single experiment to observe the time development of the stochastic variable Y. When the experiment is repeated, a different set of observations is obtained; those may be labeled by a different value of ω. Each such sequence of observed Y-values is called a realization (or sometimes a path) of the stochastic process, and from this perspective ω may be considered as labeling the realizations of the process. It is seen that it is somewhat arbitrary which of ω and t is considered to be a label, and which an independent variable; this is sometimes expressed by writing the stochastic process as $Y(t,\omega)$. In the light of these concepts and definitions we return to a consideration of the noise term W_t in equation (2.4). One can make a number of plausible assumptions about the properties ascribed to W_t, which provide us with a framework to model the noise term:

- Considering that the noise represents deviations away from the deterministic rate, the expected value of W_t over all realizations, is zero $(E(W_t) = 0)$, i.e. the mean of the noise term at a given instant is negligible.

- W_t at a particular time does not depend on the noise at any other time, i.e., W_t is an independent stochastic process.

- W_t is stationary, i.e. the joint probability distribution of $\{ W_{t1}, W_{t2}, \ldots W_{ti}, \ldots\}$ is time invariant. Unless we have a large set of "noise" data we cannot be certain about this assumption. But many applications suggest that this assumption is reasonable. If the system components and/or processes causing noise are invariant, then one could expect this assumption to hold.

- All paths of W_t are continuous. This is needed to ensure that the underlying variable, i.e. $\alpha(t)$ in equation (2.1), remains continuous within any single realization. Although the sources of noise are assumed unknown and have different values in each realization, it is reasonable to assume that they are finite and thus can not cause discontinuous jumps in any system variable.

These assumptions, even though appearing reasonable, cannot be met by any known stochastic process. A stochastic process meeting the first three requirements, cannot have continuous paths (Øksendal, 1998). Intuitively, the problem is that if at every moment the stochastic variable changes independently, any given experiment is bound to show discontinuous fluctuations of the measured value, no matter how small the time steps are made.

The resolution of this dilemma is to assume that not the variable itself, but rather its increments, are independent at different times. This state of affairs is reminiscent of the physical phenomenon of Brownian motion. The botanist Robert Brown first observed that pollen grains suspended in liquid, undergo irregular motion. Centuries later, it was realized that the physical explanation of this is that the pollen grain is continually bombarded by molecules of the liquid traveling with different speeds in different directions. Over a time scale that is large compared with the intervals between molecular impacts, these will average out and no net force is exerted on the grain. However, this will not happen over a small time interval; and if the mass of the grain is small enough to undergo appreciable displacement in the small time interval as the result of molecular impacts, an observable erratic motion results. The crucial point to notice in the present context is that while the impacts and therefore the individual displacements suffered by the grain can be considered independent at different times, the actual position of the grain can only change continuously. Clearly this phenomenon can serve as a template on which to model the description of the stochastic noise term and its mathematical description should be understood in more detail.

Let us examine the solution to the stochastic differential equation we have discussed before. We take equation (2.4) in the simple case where the population growth rate and noise amplitude are constant, i.e. $r(t) = r$ and $\sigma(t) = \sigma$. The solution to the stochastic differential equation is given by (Øksendal, 1998):

$$X_t = X_0 \exp\left(\sigma W_t\right) \exp\left(\left(r - \frac{\sigma^2}{2}\right)t\right). \tag{2.5}$$

Equation (2.5) gives a realization of the solution and each "experiment" will produce a different path, since W_t will assume a different set of values.

Comparing the stochastic solution with the deterministic solution of equation (2.2), we notice two important differences. The last exponential factor in equation (2.5) shows that the presence of noise in the rate coefficient α in the population growth equation (2.1) modifies the effective growth rate. But moreover, the first exponential factor that will develop differently for each realization, could mask completely the underlying deterministic growth if $\sigma >> r$. This also suggests that apparently random series of data may come from a simple deterministic relationship with the presence of noise in its coefficients. These remarks will be further explored by computer simulation later.

To summarize, the effects of including random effects in differential equations can manifest in two different ways which require two fundamentally different methods of analysis. The first type of equation has random coefficients, a random initial value or is forced by a fairly regular stochastic process, or when some combination of these holds. These equations are called random differential equations and are solved sample path by sample path (Kloeden and Platen, 1992). On the other hand, if the underlying noise is modeled by an irregular stochastic process such as Gaussian white noise, the equations are described as stochastic differential equations (SDE) which should be solved by employing Ito or Stratonovich stochastic integrals. We will discuss Ito integral later in this chapter and show how it differs from the standard integral, Riemann integral. The sample paths of random differential equations are differentiable but those of SDEs are not differentiable. The choice of an appropriate model to represent noise is a crucial factor in deciding the solution methods. However, most variability in natural systems appears to be irregular whether the sources of noise be the random inputs, irregular boundary conditions or unpredictable distribution of material properties. Therefore flows and variables in most natural systems can be represented as stochastic variables, and deterministic descriptions can be considered as a subset of the pertinent stochastic models.

Before we develop models using stochastic approaches, it is useful to summarily present useful concepts and methods of stochastic calculus. This discussion is brief and intended only to make reader familiar with the ideas and apply them in appropriate situations.

2.2 Elementary Stochastic Calculus

The review presented in the sections that follows this section should not be considered as an exhaustive treatment of the subject; we carefully selected concepts and definitions that are fundamental to understand stochastic calculus and processes without being overly concerned about covering the entire domain of knowledge that could be useful in developing models. Instead we present a starting point to study this area, often perceived as a difficult one in applied mathematics by the scientists and engineers who are oriented towards solving practical problems. There are many excellent books available to reader to study into the subject deeper: Klebaner (1998) gives an excellent introduction to stochastic calculus; Øksendal (1998), Kloeden and Platen (1992), Steele (2001) and Durrett (1996) provide rigorous treatments of the subject for the more mathematically oriented readers. We intend to discuss sufficient number of topics to facilitate the developments and discussions we are going to have later in the book.

2.3 What is Stochastic Calculus?

In standard calculus we deal with differentiable functions which are continuous except perhaps in certain locations of the domain under consideration. To understand the continuity of the functions better we make use of the definitions of the limits. We call a function *f*, a continuous function at the point $t = t_0$ if

$$\lim_{t \to t_0} f(t) = f(t_0)$$

regardless of the direction t approaches t_0. A right-continuous function at t_0 has a limiting value only when t approaches t_0 from the right direction, i.e. t is larger than t_0 in the vicinity of t_0. We will denote this as

$$f(t+) = \lim_{t \downarrow t_0} f(t) = f(t_0) .$$

Similarly a left-continuous function at t_0 can be represented as

$$f(t-) = \lim_{t \uparrow t_0} f(t) = f(t_0) .$$

These statements imply that a continuous function in both right-continuous and left-continuous at a given point of t. Often we encounter functions having discontinuities; hence the need for the above definitions. To measure the size of a discontinuity, we define the term "jump" at any point t to be a discontinuity where the both $f(t+)$ and $f(t-)$ exist and the size of the jump be $\Delta f(t) = f(t+) - f(t-)$. The jumps are the discontinuities of the first kind and any other discontinuity is called a discontinuity of the second kind. Obviously a function can only have countable number of jumps in a given range. From the mean value theorem in calculus it can be shown that we can differentiate a function in a given interval only if the function is either continuous or has a discontinuity of the second kind during the interval. Stochastic calculus is the calculus dealing with often non-differentiable functions having jumps without discontinuities of the second kind. One such example of a function is the Wiener process (Brownian motion). One realization of the standard Wiener process is given in Figure 2.1.

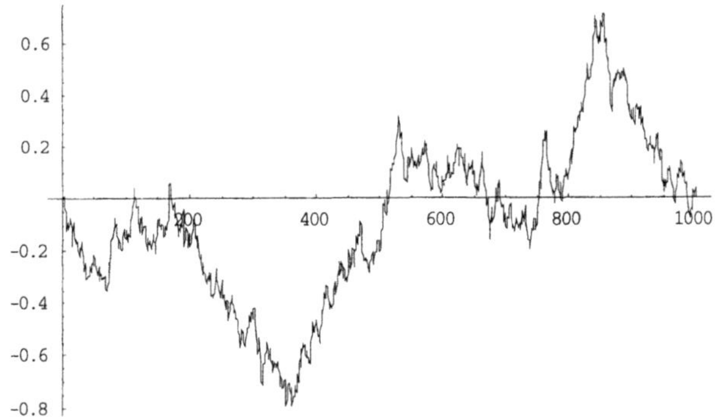

Figure 2.1 An example of a function dealt in stochastic calculus. This function is continuous but not differentiable at any point.

Without going into details of how we computed this function- we will do that in Chapter 3 – we can see that the increments are irregular and we can not define a derivative according to the mean value theorem. This is because of the fact that the function changes erratically within small intervals, however small that interval may be, and we can not define a derivative at a given point in the conventional sense. Therefore we have to devise new mathematical tools that would be useful in dealing with these irregular non-differentiable functions.

2.4 Variation of a Function

Variation of a function f on $[a,b]$ is defined as

$$V_f([a,b]) = \lim_{\delta_n \to 0} \sum_{i=1}^{n} \left| f(t_i^n) - f(t_{i-1}^n) \right|, \tag{2.6}$$

where $\delta_n = \max_{1 \le i \le n} (t_i - t_{i-1})$.

If $V_f([a,b])$ is finite such as in continuous differentiable functions then f is called a function of finite variation on $[a,b]$. Variation of a function is a measure of the total change in the function value within the interval considered. An important result (Theorem 1.7, Klebaner (1998)) is that a function of finite variation can only have a countable number of jumps. Furthermore, if f is a continuous function, f' exists and $\int |f'(t)| dt < \infty$ then f is a function of finite variation. This implies that a function of finite variation on $[a,b]$ is differentiable on $[a,b]$.

Another quantity that plays a major role in stochastic calculus is the quadratic variation. In stochastic calculus the quadratic variation of a function f over the interval $[0,t]$ is given by

$$[f](t) = \lim \sum_{i=1}^{n} (g(t_i^n) - g(t_{i-1}^n))^2 , \tag{2.7}$$

where the limit is taken over the partitions:

$$0 = t_0^n < t_1^n < \ldots < t_n^n = t ,$$

with $\quad \delta_n = \max_{1 \le i \le n} (t_i^n - t_{i-1}^n) \to 0.$

It can be proved that quadratic variation of a continuous function with finite variation is zero. However, the functions having zero quadratic variation may have infinite variation such as zero energy processes (Klebaner, 1998). If a function or process has a finite positive quadratic variation within an interval, then its variation is infinite. It also means that such functions are continuous but not differentiable.

We do not use quadratic variation in standard calculus which deals with continuous and smooth functions. We also define quadratic covariation of functions f and g on $[0, t]$ by extending equation (2.7):

$$[f, g](t) = \lim \sum_{i=0}^{n-1} (f(t_{i+1}^n) - f(t_i^n))^2 (g(t_{i+1}^n) - g(t_i^n))^2 . \tag{2.8}$$

when the limit is taken over partitions $\{t_i^n\}$ of $[0,t]$ with $\delta_n = \max_{1 \le i \le n} (t_{i+1}^n - t_i^n) \to 0.$

It can be shown that if both the functions are continuous and one is of finite variation, the quadratic covariation is zero.

Covariation of two functions, f and g, has the following properties:

Polarization Identity

Polarization identity expresses the quadratic covariation, $[f, g](t)$, in terms of quadratic variation of individual functions.

$$[f,g](t)=\frac{1}{2}([f+g,f+g](t)-[f,f](t)-[g,g](t)). \tag{2.9}$$

Symmetry

$$[f,g](t) = [g,f](t). \tag{2.10}$$

Linearity

Using polarization identity and symmetry one can show that covariation is linear for any constants a and b,

$$[af+bg,h](t)=a[f,h](t)+b[g,h](t). \tag{2.11}$$

Quadratic variation of a function $[f](t)$ and covariation $[f, g](t)$ are measures of change in the functional values over a given range $[0, t]$.

2.5 Convergence of Stochastic Processes

In many situations where stochastic processes are involved, we would like to know the limiting values of useful random variables, i.e. whether they approach some sort of a "steady state" or asymptotic behavior. However the steady state of random variables have to be defined within a probabilistic context. Therefore, in stochastic processes we discuss the convergence of random variables using four different criteria.

Almost Sure Convergence

Random variables $\{X_n\}$ converges to $\{X\}$ with probability one:

$$P(\{\omega \in \Omega : \lim_{n\to\infty} |X_n(\omega) - X(\omega)| = 0\}) = 1. \tag{2.12}$$

Mean-square Convergence

$\{X_n\}$ converges to $\{X\}$ such a way that $E(X_n^2) < \infty$ for $n = 1,2,...,n$, $E(X) < \infty$ and

$$\lim_{n\to\infty} E(|X_n - X|^2) = 0. \tag{2.13}$$

Convergence in Probability

$\{X_n\}$ converges to $\{X\}$ with zero probability of having a difference between the two processes:

$$\lim_{n\to\infty} P(\{\omega \in \Omega ; |X_n(\omega) - X(\omega)| \geq \varepsilon) = 0, \tag{2.14}$$

for all $\varepsilon > 0$.

Convergence in probability is called stochastic convergence as well.

Convergence in Distribution

Distribution function of $\{X_n\}$ converges to that of $\{X\}$ at any point of continuity of the limiting distribution (i.e. the distribution function of $\{X\}$).

These four criteria add another dimension to our discussion of the asymptotic behavior of a process. These arguments can be extended to the comparison of stochastic processes with each other. For example, in many instances one stochastic process is said to be equal to another in convergence in probability. That means for the two stochastic processes $\{X_1\}$ and $\{X_2\}$,

$$\lim_{n\to\infty} P(\{\omega\in\Omega; \left|X_1(\omega)-X_2(\omega)\right|\geq\varepsilon) = 0 \tag{2.15}$$

for all $\varepsilon>0$.

Unlike in deterministic variables where any asymptotic behavior can clearly be identified either graphically or numerically, stochastic variables do require adherence to one of the convergence criteria mentioned above which are called the "criteria for strong convergence". There are weakly converging stochastic processes and we do not discuss the weak convergence criteria as they are not relevant to the development of the material in this book.

2.6 Riemann and Stieltjes Integrals

In standard calculus we have continuous functions with discontinuities at finitely many points and we integrate them using the definition of Riemann integral of a function $f(t)$ over the interval $[a,b]$:

$$\int_a^b f(t)\,dt = \lim_{\delta\to 0}\sum_{i=1}^{n} f(\xi_i^n)\,(t_i^n - t_{i-1}^n), \tag{2.16}$$

where t_i^n 's represents partitions of the interval,

$$a = t_0^n < t_1^n < t_2^n \ldots . < t_n^n = b,$$

$$\delta = \max_{1\leq i\leq n}(t_i^n - t_{i-1}^n), \text{ and } \quad t_{i-1}^n \leq \xi_i^n \leq t_i^n .$$

Riemann integral is used extensively in standard calculus where continuous functions are the main concern. The integral converges regardless of the chosen ξ_i^n within $[t_{i-1}^n, t_i^n]$.

A generalization of Riemann integral is Stieltjes integral which is defined as the integral of $f(t)$ with respect to a monotone function $g(t)$ over the interval $[a,b]$:

$$\int_a^b f(t)\,dg(t) = \lim_{\delta \to 0} \sum_{i=1}^{n} f(\xi_i)(g(t_i^n) - g(t_{i-1}^n)), \tag{2.17}$$

with the same definitions as above for δ and t_i^n's. It can be shown that for the integral to exist for any continuous function $f(t)$, $g(t)$ must be a function with finite variation on $[a,b]$. This means that if $g(t)$ has infinite variation on $[a,b]$ then for such a function, integration has to be defined differently. This is the case in stochastic integration based on Brownian motion which has infinite variation in any given interval however small the interval is, and therefore, can not be integrated using Stieltjes integral. Before we discuss alternative forms of integration that can be applied to the functions of positive quadratic variation, i.e. the functions of infinite variation, we will present important properties and some useful results associated with Brownian motion or the Wiener process.

2.7 Brownian Motion and Wiener Processes

In the physical Brownian motion, there are small but nevertheless finite intervals between the impulses of molecules colliding with the pollen grain. Consequently, the path that the grain follows, consists of a sequence of straight segments forming an irregular but continuous line – a so-called random walk. Each straight segment can be considered an increment of the momentary position of the grain.

The mathematical idealization of this, similar to that which describes a circle as the limiting case of a polygon when the number of sides approaches infinity, is to let the interval between increments approach zero. The resulting process – called Brownian motion or a Wiener process – is difficult to conceptualize: for example, consideration shows that the resulting position is everywhere continuous, but nowhere differentiable. This means that while the particle has a position at any moment, and since this position is changing - it is moving - yet no velocity can be defined. Nevertheless as discussed by Stroock and Varadhan (1979) a consistent mathematical description is obtained by defining the position as a stochastic process $B(t,\omega)$ with the following properties that are suggested as a model for a pure noise process:

P1: $B(0,\omega) = 0$, i.e. choose the position of the particle at the arbitrarily chosen initial time $t = 0$ as the coordinate origin;

P2: $B(t,\omega)$ has independent increments, i.e. $B(t_1,\omega)$, $\{B(t_2,\omega)- B(t_1,\omega)\},\ldots,$ $\{B(t_k,\omega)- B(t_{k-1},\omega)\}$ are independent for all $0 \leq t_1 < t_2 \ldots < t_k$;

P3: $\{B(t_{i+1},\omega)-B(t_i,\omega)\}$ is normally distributed with mean 0 and variance $(t_{i+1} - t_i)$;

P4: The stochastic variation of $B(t,\omega)$ at fixed time t is determined by a Gaussian probability;

P5: The Gaussian has a zero mean, $E[B(t,\omega)] = 0$ for all values of t;

P6: $B(t,\omega)$ are continuous functions of t for $t \geq 0$;

P7: The covariance of Brownian motion is determined by a correlation between the values of $B(t,\omega)$ at times t_i and t_j (for fixed ω), given by

$$E[\, B(t_i,\omega)\, B(t_j,\omega)\,] = \min(t_i,t_j). \tag{2.18}$$

When applied to $t_i = t_j = t$, P7 reduces to the statement that

$$Var[B(t,\omega)] = t, \tag{2.19}$$

where '*Var*' means statistical variance. For Brownian motion this can be interpreted as the statement that the radius within which the particle can be found increases proportional to time. This is a plausible behavior for a random walk phenomenon, and is of such fundamental importance in what follows it is explored in more detail.

Consider a particle restricted to one-dimensional motion along the x-axis, starting from an initial position $x=x_0$. It is acted upon by independent impacts (e.g. from gaseous molecules impinging on it) at an average rate of f impacts per unit time. Its displacement $b(\tau)$ after a time τ, is given by

$$b(\tau) = x_0 + \sum_{i=1}^{N} x_i \,, \tag{2.20}$$

where $x_i = v_i \Delta t_i$ is the distance traveled in interval i as a result of the velocity v_i it acquires in the i-th impact occurring at a discrete time t_i. In terms of the previous terminology, the x_i s are the increments of the position. The total number of impacts N is obviously given by $N = f\,\tau$. The quantities v_i and Δt_i have probability distributions which will be determined by the physics of the situation, but are not further specified except for the assumption that the average value of v_i and consequently also of x_i, are zero. Considering each x_i to be an independent stochastic variable, the probability distribution of $b(\tau)$ is determined by the so-called Central Limit Theorem (CLT) of elementary statistics (Kenney 1966, or any standard statistics textbook). According to the CLT, the distribution of a sum of stochastic variables approaches a normal (i.e. Gaussian) distribution, with its mean and variance equal to the sum of means and variances of the individual variables, as the number of terms approaches infinity. This applies for any non-pathological distribution of the individual variables. Hence $b(\tau)$ has a Gaussian probability distribution with zero mean, and its variance is N times that of an individual position increment x_i.

For a fixed average impact frequency, this means that $Var(b) \propto \tau$ as long as $\tau >> 1/f$ so that $N >> 1$. On the other hand, suppose we keep τ fixed and let f increase without changing the distribution of the impact velocities. For example, in the actual experiment the density of the gas may be increased without changing the temperature. Then, although N increases proportional to f, the value and therefore also the variance of each x_i decreases in the same ratio because the Δt_i decreases proportional to $1/f$. Therefore it is reasonable to assume $Var(b)$ is independent of f and we can take it as proportional to τ even in the limit as $f \to \infty$, in which case the discrete step Brownian motion becomes a Wiener process.

In this way the set of Wiener process properties stipulated above are seen to arise naturally from consideration of a random walk. In particular, the assumption of a Gaussian distribution for B is relatively independent of the detailed statistical properties of the increments.

Note that in the Brownian motion example, τ is multiplied by a proportionality constant containing the average impact frequency and the variance of individual increments, but in the Wiener process the time constant is one. To achieve that in the Brownian motion, either the position variable or the time needs to be appropriately rescaled. In adopting the standard Wiener process definition, this scaling has been hidden from view. As often done in mathematical discussions, all variables are essentially assumed to be dimensionless. This convention needs to be remembered when applying the theory to a physical situation.

A consistent way to do this is to transform all physical variables occurring in the applicable deterministic differential equations to dimensionless ratios, by dividing them by appropriate scale constants, before introducing the stochastic terms to the equation. In choosing scales one should recognize that the Wiener definition itself introduces the new scale constant explained above. In our Brownian motion example, the rate at which the particle wanders away from its starting position will clearly depend on the magnitude of the velocity imparted to it in individual impacts, i.e. on the mass of the particle and the temperature of the gas in which it is immersed. This demonstrates that physical stochastic processes can take place on different time scales, and an appropriate one should be used to reduce a particular problem to the universal time scale assumed for a Wiener process.

In the previous discussion, for the sake of clarity a distinction was made between Brownian motion where there are random increments at discrete time steps, and the Wiener process which is the limit in which the intervals between increments approach zero. Many authors do not make this distinction and use the terms Brownian motion and Wiener process interchangeably for the mathematical idealization. We will also use the terms Brownian motion and Wiener process interchangeably and by doing so we refer to the same stochastic process.

Because the Wiener process is defined by the independence of its increments, it is for some purposes convenient to reformulate the variance stipulation of a Wiener process in terms of the variance of the increments:

From P3, for $t_i < t_j$:

$$\mathrm{var}[B(t_j,\omega)-B(t_i,\omega)] = t_j - t_i \ . \tag{2.21}$$

Bearing in mind that the statistical definition of the variance of a quantity X reduces to the expectation value expression $Var(X) = E(X^2)\text{-}E^2(X)$ and that the expectation value or mean of a Wiener process is zero, we can rewrite this as

$$E[\{B(t_2,\omega)-B(t_1,\omega)\}^2]=\text{var}[B(t_2,\omega)-B(t_1,\omega)]$$

i.e. $$E[\Delta B\cdot\Delta B]=\Delta t \tag{2.22}$$

where Δ is defined to mean the time increment for a fixed realization ω. The connection between the two formulations is established by similarly rewriting equation (2.21) and then applying equation (2.18):

$$\begin{aligned} Var[B(t_1,\omega)-B(t_2,\omega)] &= E[\{B(t_1,\omega)-B(t_2,\omega)\}^2] \\ &= E[B^2(t_1,\omega)+B^2(t_1,\omega)-2B(t_1,\omega)B(t_2,\omega)] \\ &= t_1+t_2-2\min(t_1,t_2) \\ &= t_1-t_2 \qquad \text{for } t_1>t_2\,. \end{aligned}$$

2.8 Relationship between White Noise and Brownian Motion

Consider a stochastic process $X(t,\omega)$ having a stationary joint probability distribution and $E(X(t,\omega))=0$, i.e. the mean value of the process is zero. The Fourier transform of $Var(X(t,\omega))$ can be written as,

$$S(\lambda,\omega)=\frac{1}{2\pi}\int_{-\infty}^{\infty} Var(X(\tau,\omega)\,e^{-i\lambda\tau}d\tau \tag{2.23}$$

$S(\lambda,\omega)$ is called the spectral density of the process $X(t,\omega)$ and is also a function of angular frequency λ. The inverse of the Fourier transform is given by

$$Var(X(\tau,\omega))=\int_{-\infty}^{\infty} S(\lambda,\omega)\,e^{i\lambda\tau}d\lambda \tag{2.24}$$

and when $\tau=0$,

$$Var(X(0,\omega))=\int_{-\infty}^{\infty} S(\lambda,\omega)d\lambda\,. \tag{2.25}$$

Therefore, variance of $X(0,\omega)$ is the area under a graph of spectral density $S(\lambda,\omega)$ against λ:

$$Var(X(0,\omega)) = E(X^2(0,\omega)), \tag{2.26}$$

because $E(X(t,\omega))=0$.

Spectral density $S(\lambda,\omega)$ is considered as the "average power" per unit frequency at λ which gives rise to the variance of $X(t,\omega)$ at $\tau = o$. If the average power is a constant which means that the power is distributed uniformly across the frequency spectrum, such as the case for white light, then $X(t,\omega)$ is called white noise. White noise is often used to model independent random disturbances in engineering systems, and the increments of Brownian motion have the same characteristics as white noise. Therefore white noise $(\zeta(t))$ is defined as

$$\zeta(t) = \frac{dB(t)}{dt},$$

$$dB(t) = \zeta(t)dt. \tag{2.27}$$

We will use this relationship to formulate stochastic differential equations.

2.9 Relationships Among Properties of Brownian Motion

As shown before, the relationships among the properties mentioned above can be derived starting from P1 to P7. For example, let us evaluate the covariance of Brownian motions of $B(t_i,\omega)$ and $B(t_j,\omega)$:

$$Cov(B(t_i,\omega)B(t_j,\omega)) = E(B(t_i,\omega)\, B(t_j,\omega)). \tag{2.28}$$

Assuming $t_i < t_j$ we can express

$$B(t_j,\omega) = B(t_i,\omega) + B(t_j,\omega) - B(t_i,\omega). \tag{2.29}$$

Therefore,

$$E(B(t_i,\omega)\, B(t_j,\omega)) = E(B(t_i,\omega)(B(t_i,\omega) + B(t_j,\omega) - B(t_i,\omega)),$$

$$= E(B^2(t_i,\omega) + B(t_i,\omega)B(t_j,\omega) - B^2(t_i,\omega)),$$

$$= E(B^2(t_i,\omega) + B(t_i,\omega)(B(t_j,\omega) - B(t_i,\omega))),$$

$$= E(B^2(t_i,\omega)) + E(B(t_i,\omega)(B(t_j,\omega) - B(t_i,\omega))). \tag{2.30}$$

From P2, $B(t_i,\omega)$ and $(B(t_j,\omega) - B(t_i,\omega))$ are independent processes and therefore we can write

$$E(B(t_i,\omega)(B(t_j,\omega) - B(t_i,\omega))) = E(B(t_i,\omega))E(B(t_j,\omega) - B(t_i,\omega)). \tag{2.31}$$

According to P3 and P5,

$E(B(t_i,\omega)) = 0$ and

$E(B(t_j,\omega) - B(t_i,\omega) = 0$.

Therefore, from equation (2.31)

$E(B(t_i,\omega)B(t_j,\omega) - B(t_i,\omega))) = 0$.

This leads equation (2.30) to

$E(B(t_i,\omega)B(t_j,\omega)) = E(B^2(t_i,\omega))$, and

$$E(B^2(t_i,\omega)) = E((B(t_i,\omega)) - 0)^2). \tag{2.32}$$

From P3, { $B(t_i,\omega) - B(0,\omega)$ } is normally distributed with a variance $(t_i - 0)$, and equation (2.32) becomes,

$$E(B^2(t_i,\omega)) = t_i \tag{2.33}$$

and, therefore,

$$Cov(B(t_i,\omega)B(t_j,\omega)) = t_i. \tag{2.34}$$

Using a similar approach it can be shown that if $t_i > t_j$,

$$Cov(B(t_i,\omega)B(t_j,\omega)) = t_j. \tag{2.35}$$

This leads to P7:

$$E(B(t_i,\omega)B(t_j,\omega)) = \min(t_i,t_j). \tag{2.36}$$

The above derivations show the relatedness of the variance of an independent increment, $Var\{B(t_1,\omega)-B(t_2,\omega)\}$ to the properties of Brownian motion given by P1 to P7. The fact that $\{B(t_{i+1},\omega)-B(t_i,\omega)\}$ is a Gaussian random variable with zero mean and $\{t_{i+1}-t_i\}$ variance can be used to construct Brownian motion paths on computer. If we decide the time interval $[0,t]$ into n equidistant parts having length Δt, and at the end of each segment we can randomly generate a Brownian increment using the Normal distribution with mean 0 and variance Δt. This increment is simply added to the value of Brownian motion at the point considered and move on to the next point. When we repeat this procedure starting, from $t=\Delta t$ to t=t and taking the fact that $B(0,\omega)=0$ into account, we can generate a realization of Brownian motion. We can expect these Brownian motion realizations to have properties quite distinct from other continuous functions of t. We will briefly discuss some important characteristics of Brownian motion realizations next as these results enable us to utilize this very useful stochastic process effectively.

2.10 Further Characteristics of Brownian Motion Realizations

1. *$B(t,\omega)$ is a continuous, nondifferentiable function of t.*

2. *The quadratic variation of $B(t,\omega)$, $[B(t,\omega),B(t,\omega)](t)$ over $[0,t]$ is t.*

Using the definition of covariation of functions,

$$\begin{aligned}[B(t,\omega),B(t,\omega)](t) &= [B(t,\omega),B(t,\omega)]([0,t]) \\ &= \lim_{\delta_n\to 0}\sum_{i=1}^{n}[B(t_i^n)-B(t_{i-1}^n)]^2\end{aligned} \tag{2.37}$$

where $\delta_n=\max(t_{i+1}^n-t_i^n)$ and $\{t_i^n\}_{i=1}^n$ is a partition of $[0,t]$, as $n\to\infty,\delta_n\to 0$.

Taking the expectation of the summation,

$$E(\sum(B(t_i^n)-B(t_{i-1}^n))^2)=\sum(E((B(t_i^n)-B(t_{i-1}^n))^2)) \tag{2.38}$$

$E((B(t_i^n)-B(t_{i-1}^n))^2)$ is the variance of an independent increment $\{B(t_i^n)-B(t_{i-1}^n)\}$.

As seen before,

$$Var[B(t_i^n)-B(t_{i-1}^n)]=(t_i^n-t_{i-1}^n). \tag{2.39}$$

Therefore,

$$\begin{aligned} E\left(\sum(B(t_i^n)-B(t_{i-1}^n))^2\right) &= \sum Var[B(t_i^n)-B(t_{i-1}^n)], \\ &= \sum_{i=1}^{n}(t_i^n-t_{i-1}^n)=t-0=t. \end{aligned} \tag{2.40}$$

Let us take the variance of $\sum(B(t_i^n)-B(t_{i-1}^n))^2$:

$$Var\left(\sum(B(t_i^n)-B(t_{i-1}^n))^2\right)=\sum 3(t_i^n-t_{i-1}^n)^2 \le 3\ \max\,(t_i^n-t_{i-1}^n)\,t=3t\delta_n. \tag{2.41}$$

As $n\to\infty, \delta_n\to 0, \sum Var(B(t_i^n)-B(t_{i-1}^n))^2\to 0$.

Summarizing the results,

$$E(\sum(B(t_i^n)-B(t_{i-1}^n))^2)=t$$

and

$$Var(\sum(B(t_i^n)-B(t_{i-1}^n))^2)\to 0 \ \text{as} \ n\to\infty.$$

This implies that, according to the monotone convergence theories that $\sum(B(t_i^n)-B(t_{i-1}^n))^2\to t$ almost surely as $n\to\infty$.

Therefore, the quadratic variation of Brownian motion $B(t,\omega)$ is t:

$$[B(t,\omega),B(t,\omega)](t)=t. \tag{2.42}$$

Omitting t and ω, $[B,B](t)=t$.

3. Brownian Motion $(B(t,\omega))$ is a Martingale.

A stochastic process, $\{X(t)\}$ is a martingale, when the future expected value of $\{X(t)\}$ is equal to $\{X(t)\}$. In mathematical the notation, $E(X(t+s)|F_t) = X(t)$ with converging almost surely, and F_t is the information about $\{X(t)\}$ up to time t. We do not give the proof of these martingale characteristics of Brownian motion here but it is easy to show that $E(B(t+s)\,|\,F_t) = B(t)$.

It can also be shown that $\{B(t,\omega)^2 - t\}$ and $\{\exp(\alpha B(t,\omega) - \frac{\alpha^2}{2}t)\}$ are also martingales. These martingales can be used to characterize Brownian motion as well and more details can be found in Klebaner (1998).

4. Brownian motion has Markov property.

Markov property simply states that the future of a process depends only on the present state. In other words, a stochastic process having Markov property does not "remember" the past and the present state contains all the information required to drive the process into the future states.

This can be expressed as

$$P(X(t+s) \le y\,|\,F_t) = P(X(t+s) \le y\,|\,X(t)), \tag{2.43}$$

almost surely .

From the very definition of increments of the Wiener process (Brownian motion), for the discretized intervals of [0,t], $\{B(t_{i+1}^n) - B(t_i^n)\}$, the Brownian motion increment behaves independently to its immediate processor $\{B(t_i^n) - B(t_{i-1}^n)\}$. In other words $\{B(t_{i+1}^n) - B(t_i^n)\}$ does not remember the behavior of $\{B(t_{i-1}^n) - B(t_{i-1}^n)\}$ and only element common to both increments is $B(t_i^n)$. One can now see intuitively why Brownian motion should behave as a Markov process. This can be expressed as

$$P(B(t_i + s) \le x_i\,|\,\{B(t_i), B(t_{i-1})...0)\}) = P(B(t_i + s) \le x_i\,|\,B(t_i)), \tag{2.44}$$

which is another way of expressing the previous equation (2.43).

2.11 Generalized Brownian Motion

The Wiener process as defined above is sometimes called the standard Wiener process, to distinguish it from that obtained by the following generalized equation (2.45):

$$E[\ B(t_i,\omega)\ B(t_j,\omega)\ ] = \int_0^{\min(t_i,t_j)} q(\tau)\,d\tau \ . \tag{2.45}$$

The integral kernel $q(\tau)$ is called the correlation function and determines the correlation between stochastic process values at different times. The standard Wiener process is the simple case that $q(\tau) \equiv 1$, i.e. full correlation over any time interval; the generalized Wiener process includes, for example, the case that q decreases, and there is progressively less correlation between stochastic values in a given realization as the time interval between them increases.

2.12 Ito Integral

At this point of our discussion, we need to define the integration of stochastic process with respect to the Wiener process $(B(t,\omega))$ so that we understand the conditions under which this integral exists and what kind of processes can be integrated using this integral. As we restrict the definition to Ito integration we denote the integral as

$$I[X](\omega) = \int_S^T X(t,\omega)\,dB(t,\omega)\,. \tag{2.46}$$

$I[X](\omega)$ implies that the integration of $X[t,\omega]$ is along a realization ω and with respect to Brownian motion which is a function of t. $I[X](\omega)$ is also a stochastic process in its own right and have properties stemming out of the definition of the integral. It is natural to expect $I[X](\omega)$ to be equal to $c(B(t,\omega) - B(s,\omega))$ when $X(t,\omega)$ is a constant c. If $X(\ t\ )$ is a deterministic process, which can be expressed as a sequence of constants over small intervals, we can define Ito integral as follows:

$$\begin{aligned} I[X] &= \int_S^T X(t)\,dB(t) \\ &= \sum_{i=0}^{n-1} c_i((B(t_{i+1}) - B(t_i))) \end{aligned} \tag{2.47}$$

where $X(t)=\begin{cases} c_0, & t=S \\ c_i, & t_i<t\leq t_{i+1} \end{cases}$ $i=0,\cdots,n-1$.

The time interval $[S,T]$ has been discretized into n intervals: $S=t_0<t_1<\cdots<t_n=T$.

Using the property of independent increments of Brownian motion, we can show that the mean of $I[X](\omega)$ is zero and,

$$Variance = Var(I[X])$$
$$= \sum_{i=0}^{n-1} c_i^2 (t_{i+1}-t_i). \tag{2.48}$$

It turns out that if $X(t,\omega)$ is a continuous stochastic process and its future values are solely dependant on the information of this process only up to *t*, Ito integral $I[X](\omega)$ exists. The future states of a stochastic process, $X(t,\omega)$, is only dependent on F_t then it is called an adapted process. A left-continuous adapted process $X(t,\omega)$ is defined as a predictable process and it satisfies the following condition: $\int_0^T X^2(t,\omega)\,dt<\infty$ almost surely.

As we are only concerned about continuous processes driven by the past events, we limit our discussion of predictable processes to the subclass of left-continuous and adapted processes. Reader may want to refer to Øksendal (1998) and Klebaner (1998) for more rigorous discussion of these concepts.

We can now define Ito integral $I[X](\omega)$ similarly to equation (2.47). If $X(t,\omega)$ is a continuous and adapted process then $I[X](\omega)$ can be defined as

$$\sum_{i=0}^{n-1} X(t_i^n,\omega)(B(t_{i+1}^n,\omega)-B(t_i^n,\omega)), \tag{2.49}$$

and this sum converges in probability.

Dropping ω for convenience and adhering to the same discretization of interval [*S*, *T*] as in equation (2.47),

$$I[X]=\int_S^T X(t)dB(t)=\sum_{i=0}^{n-1} X(t_i^n)(B(t_{i+1}^n)-B(t_i^n)). \tag{2.50}$$

Equation (2.50) expresses an approximation of $\int_S^T X(t)\,dB(t)$ based on probabilistic convergence. We take equation (2.49) as the definition of Ito integral for the purpose of this book. As stated earlier $I[X](\omega)$ is a stochastic process and it has the following properties (see, for example, Øksendal (1998) for more details):

1. Ito integral is linear.

If $X(t)$ and $Y(t)$ are predictable processes and α and β as some constants, then

$$I[\alpha X + \beta Y](\omega) = \alpha\, I[X](\omega) + \beta\, I[Y](\omega)\,. \tag{2.51}$$

2. Ito integral has zero-mean property.

$$E(I[X](\omega)) = 0\,. \tag{2.52}$$

3. Ito integral is isometric.

$$E[(\int_S^T X(t)\,dB(t))^2] = \int_S^T E(X^2(t))\,dt\,. \tag{2.53}$$

The isometry property says that the expected value of the square of Ito integral is the integral with respect to t of the expectation of the square of the process $X(t)$.

Since $E[(\int_S^T X(t)\,dB(t))^2]$=0 from zero mean property, we can express the left hand side of equation (2.53) as

$$\begin{aligned} &E((\int_S^T X(t)dB(t))^2 - E(\int_S^T X(t)dB(t))) \\ &= E[\int_S^T X(t)dB(t) - E(\int_S^T X(t)dB(t))]^2 = Var\left(\int_S^T X(t)dB(t)\right). \end{aligned} \tag{2.54}$$

Therefore the variance of Ito integral process is $\int_s^T E(X^2(t))dt$ and this result will be useful to us in understanding the behavior of Ito integral process. We say that Ito integral is square integrable. According to Fubuni's Theorem, which states that, for a stochastic process $X(t)$, with continuous realizations,

$$E(\int_s^T X(t)dt) = \int_s^T E(X(t))dt, \tag{2.55}$$

$$\int_s^T E(X^2(t))dt = E(\int_s^T (X^2(t)dt). \tag{2.56}$$

4. *Ito integral is a martingale.*

It can be shown that $E(I[X(t)] \mid F_t) = I[X(t)]$. Strictly speaking $X(t)$ should satisfy $\int_S^T X^2(t)dt < \infty$ and $\int_S^T E(X^2(t))dt < \infty$ for martingale property to be true. Therefore Ito integrals are square integrable martingales.

5. *Ito integral of a deterministic function* X (t) *is a Guassian process with zero mean and covariance function.*

$$Cov(I[X(t)], I[X(t+t_0)]) = \int_S^t X^2(s)ds, \ t_0 \geq 0. \tag{2.57}$$

$I\,[X\,(t\,)\,]$ is a square integrable martingale.

6. *Quadratic variation of Ito integral is given by*

$$[I[X], P[X]](t) = \int_S^T X^2(t)dt. \tag{2.58}$$

We see that Ito integral has a positive quadratic variation making it a process with infinite variation i.e. it is a nondifferentiable continuous function of *t*.

7. *Quadratic covariation of Ito integral with respect to processes* $X_1(t)$ *and* $X_2(t)$ *is given by*

$$[I[X_1], I[X_2]](f) = \int_S^T X_1(t)X_2(t)dt. \tag{2.59}$$

Armed with these properties we can proceed to discuss mechanics of stochastic calculus such as stochastic chain rule, which is also known as Ito formula.

2.13 Stochastic Chain Rule (Ito Formula)

2.13.1 Differential notation

As we have seen previously, quadratic variations of Brownian motion, $[B(t,\omega), B(t,\omega)](t)$, is the limit in probability over the interval $[0, t]$:

$$[B(t,\omega),B(t,\omega)](t) = \lim_{\delta_n \to 0} \sum_{i=0}^{n-1} (B(t_{i+1}^n) - B(t_i^n))^2, \tag{2.60}$$

$\delta_n = \max(t_{i+1}^n - t_i^n) \to 0$. Using the differential notation,

$$\Delta B = B(t_{i+1}^n) - B(t_i^n),$$

and summation as an integral,

$$[B(t,\omega),B(t,\omega)](t) = \int_s^t (dB(s))^2 . \tag{2.61}$$

We have shown that $[B,B](t) = t$, and therefore, $\int_s^t (dB(s))^2 = t$.

For our convenience and also to make the notation similar to the one in standard differential calculus, we denote

$$\int_o^t (dB(s))^2 = t \tag{2.62}$$

as $(dB(t))^2 = dt$. (2.63)

This equation does not have a meaning outside the integral equation (2.62) and should not be interpreted in any other way.

Similarly for any other continuous function $g(t)$,

$$g(t)(dB(t))^2 = g(B(t))dt\,, \tag{2.64}$$

which means,

$$\int_o^t g(t)dB(s))^2 = \int_o^t g(B(s))ds. \tag{2.65}$$

This equation is an expression of the approximation, in probability, of

$$\lim_{\delta_n \to 0} \sum_{i=0}^{n-1} g(t_i^n)(B(t_{i+1}^n) - B(t_i^n))^2 = \int_o^t g(B(s))ds\,. \tag{2.66}$$

As quadratic variation of a continuous and differentiable function is zero,

$$[t,t]\ (t) = 0\,. \tag{2.67}$$

This equation in integral notation,

$$\int_0^t (dt)^2 = 0\ ,$$

and in differential notation,

$$(dt)^2 = 0\ . \tag{2.68}$$

Similarly, quadratic covariation of t (a continuous and differentiable function) and Brownian notion,

$$[t,B](t) = 0\,. \tag{2.69}$$

This relationship can be proved by expressing quadratic covariation as

$$[t,B](t) = \lim_{\delta_n \to o} \sum_{i=0}^{n-1} (t_{i+1}^n - t_i^n)(B(t_{i+1}^n) - B(t_i^n)), \tag{2.70}$$

$$\delta_n = \max(t_{i+1}^n - t_i^n),$$

$$\begin{aligned} [t,B](t) &\le \delta_n \sum_{i=0}^{n-1} (B(t_{i+1}^n) - B(t_i^n)), \\ &\le \delta_n B(t). \end{aligned} \tag{2.71}$$

Therefore as $n \to \infty, \delta_n \to 0$ (because t is a function of finite variation),

$$[t, B](t) \to 0 \quad \text{as} \quad n \to \infty .$$

Hence, $[t, B](t) = 0$ and in integral notation,

$$\int_0^t dt\, dB = 0 \ . \tag{2.72}$$

This can be written in differential notation,

$$dt.dB = 0 . \tag{2.73}$$

Therefore, we can state the following rules in differential notation,

$$dt.dt = 0\,;\ dt.dB = 0\,;\ dB.dt = 0,\ \text{and}\ \ dB.dB = dt\ . \tag{2.74}$$

2.13.2 Stochastic Chain Rule

In order to come to grips with the interpretation of the differential properties of dB_t, it is useful to consider the chain rule of differentiation. This will also lead us to formulas that are often more useful to apply in calculating Ito integrals than the basic definition as the limit of a sum. Consider first the case in ordinary calculus of a function g (x,t), where x is also a function of t. We can write the change in g as t changes, as follows:

$$\Delta g = \frac{\partial g(t,x)}{\partial t}\Delta t + \frac{\partial g(t,x)}{\partial x}\Delta x + \tfrac{1}{2}\frac{\partial^2 g(t,x)}{\partial x^2}(\Delta x)^2 + ...$$

From this, an expression for dg/dt is obtained by taking the limit $\Delta t \to 0$ of the ratio $(\Delta g/\Delta t)$.

Since $\Delta x = (dx/dt)\ \Delta t$, when $\Delta t \to 0$ the 2nd derivative term shown is of order $(\Delta t)^2$ and falls away together with all higher derivatives, and the well-known chain rule formula for the total derivative (dg/dt) is obtained. However, if instead of x we have a Wiener process B_t, we get

$$\Delta g = \frac{\partial g(t,B_t)}{\partial t}\Delta t + \frac{\partial g(t,B_t)}{\partial x}\Delta B_t + \tfrac{1}{2}\frac{\partial^2 g(t,B_t)}{\partial x^2}(\Delta B_t \cdot \Delta B_t) + \ldots$$

If the expectation value of this expression over all realizations is taken, the above shows that the second derivative term is now only of order Δt and cannot be ignored. Since this holds for the expectation value, for consistency we also cannot neglect the term if the limit $\Delta t \to 0$ is taken without considering the expectation value. Unlike the case of ordinary calculus where all expressions containing products of differentials higher than 1 is neglected, in Ito calculus we therefore have the rules given by equation (2.74).

Recall that in standard calculus chain rule is applied to composite functions. For example, if $Y=f(t)$ then $g(Y)$ is a function of Y.

Then

$$\frac{dg}{dt} = \frac{dg}{dY}.\frac{dY}{dt}.$$

In differential notation,

$$dg = \frac{dg}{dY}.df .$$

By integrating

$$g(f(t)) = g(0) + \int_0^t g'(f(t))df . \tag{2.75}$$

Suppose say $f(t) = B(t)$ (Brownian motion) and $g(x)$ is twice continuously differentiable continuous function. Then by using stochastic Taylor series expansion,

$$g(B(t)) = g(0) + \int_0^t g'(B(s))dB(s) + \frac{1}{2}\int_0^t g''(B(s))ds . \tag{2.76}$$

Comparing equation (2.76) and the corresponding stochastic chain rule, we can see that the second derivative term of the Taylor series plays a significant role in changing the chain rule in the standard calculus to the stochastic one. For example, let

$$g(x) = e^x .$$

Therefore,

$$e^{B(t)} = e^{(0)} + \int_0^t e^{B(s)} dB(s) + \frac{1}{2}\int_0^t e^{B(s)} ds \ . \tag{2.77}$$

In differential notation (which is only a convention)

$$d(e^{B(t)}) = e^{B(t)} dB(t) + \frac{1}{2} e^{B(t)} dt \ . \tag{2.78}$$

As another example, let $g(x) = x^2$.

Therefore, from the chain rule

$$(B(t))^2 = (B(0))^2 + 2\int_s^t B(s)dB(s) + \frac{1}{2}\int_0^t 2ds\,, \tag{2.79}$$

$$\int_0^t B(s)dB(s) = \frac{1}{2}(B(t))^2 - \frac{1}{2}t\,. \tag{2.80}$$

This is quite a different result from the standard integration. In differential convention,

$$B(t)\, dB(t) = \frac{1}{2} d((B(t))^2) - \frac{1}{2} dt\,. \tag{2.81}$$

In other words, the stochastic process $\int_0^t B(s)dB(s)$ can be calculated by evaluating $\{\frac{1}{2}(B(t))^2 - \frac{1}{2}t\}$. We will show how this process behaves using computer simulations in Chapter 3.

2.13.4 Ito Processes

We extend Ito integral by defining on stochastic process, which is similar form to stochastic differential equation (2.4) when written in differential from. We write Ito integral as

$$Y(t) = \int_0^t \sigma(s)dB(s). \tag{2.82}$$

The we can add a "drift term" to the "diffusion term" given by equation (2.82):

$$Y(t) = Y(0) + \int_0^t \mu(s)ds + \int_0^t \sigma(s)dB(s). \tag{2.83}$$

We recall that $\sigma(s)$ should be a predictable process and is subjected to the condition $\int_0^T \sigma^2(t)dt < \infty$ almost surely. $\mu(t)$ is, on the other hand, an adapted continuous process of finite variation. In equation (2.83) $\int_0^t \sigma(s)dB(s)$ represents the diffusion part of the process and $\int_0^t \mu(s)ds$ does not contain the noise; therefore it represents the drifting of the process. $Y(t)$ is called an Ito process and in differential notation we can write,

$$dY(t) = \mu(t)dt + \sigma(t)dB(t). \tag{2.84}$$

Equation (2.84) can be quite useful in practical applications where the main driving force is perturbed by an irregular noise. A particle moving through a porous medium is such an example. In this case, advection gives rise to the drift term and hydrodynamic dispersion and micro diffusion give rise to the "diffusion" term. In the population dynamics example we previously discussed, the diffusion term is a direct result of noise in the proportionality constant. Therefore it is important to study Ito process further in order to apply it in modeling situations. $\mu(t)$ is called the drift coefficient and $\sigma(t)$ the diffusion coefficient and they can depend on $Y(t)$ and/or $B(t)$. For example, we can write in pervious result (equation (2.81)),

$$d(B(t)^2) = dt + 2B(t)dB(t). \tag{2.85}$$

This is an Ito process with the drift coefficient of 1 and the diffusion coefficient of $2B(t)$. Quadratic variation of Ito process on $[0,T]$

$$Y(t) = Y(0) + \int_0^t \mu(s)ds + \int_0^t \sigma(s)dB(s) \tag{2.86}$$

is given by

$$[Y,Y](t) = \int_0^t \sigma^2(s)ds\,. \tag{2.87}$$

This can be deduced from the fact that $\int_0^t \mu(s)ds$ is a continuous function with finite variation and using quadratic variation of Ito integral. In differential notation,

$$\begin{aligned}(dY(t))^2 &= dY(t).dY(dt),\\ &= \mu^2(t)(dt)^2 + 2\mu\,\sigma dt dB + \sigma^2(dB)^2,\\ &= \sigma^2(t)dt.\end{aligned} \tag{2.88}$$

The chain rule given in equation (2.76) gives us a way to compute the behavior of a function of Brownian motion. It is also useful to know the chain rule to compute a function of a given Ito process. Suppose Ito process is given by a general form,

$$dX(t) = \mu dt + \sigma dB(t)\,. \tag{2.89}$$

where μ is the drift coefficient and B is the diffusion coefficient and let $g\ (t,\ x)$ is a twice differentiable continuous function. Let $Y\ (t) = g(t,\ X\ (t))$. Here $Y\ (t)$ is a function of t and Ito process $X\ (t)$, and also a stochastic process. $Y\ (t)$ can also be expressed as an Ito process. Then Ito formula states,

$$dY(t) = \frac{dg}{dt}(t, X(t))dt + \frac{\partial g}{\partial x}(t, X(t))dX(t) + \frac{1}{2}\frac{\partial^2 g}{\partial x^2}(t, X(t)).(dX(t))^2 \tag{2.90}$$

where ,

$$(dX(t))^2 = d(X(t)).d(X(t))\,, \tag{2.91}$$

and is evaluated according to the rules given by equation (2.74).

As an example, consider the Ito process

$$dX(t) = dt + 2B(t)dB(t) \tag{2.92}$$

where $\mu = 1$ and $\sigma = 2B(t)$.

Assume $g(t, x) = x^2$, therefore

$$\frac{\partial g}{\partial t} = 0; \ \frac{\partial g}{\partial x} = 2x; \ \frac{\partial^2 g}{\partial^2 x} = 2. \tag{2.93}$$

Substituting to Ito formula above,

$$\begin{aligned} dg &= 2X(t)dX(t) + dX(t).dX(t), \\ &= 2(dt + 2B(t)dB(t))X(t) + 4B^2(t)dt. \end{aligned} \tag{2.94}$$

Therefore,

$$dX^2(t) = (2X(t) + 4B^2(t))dt + 4X(t)B(t)dB(t). \tag{2.95}$$

As seen above $dX^2(t)$ is also an Ito process with $u = 2X(t) + 4B^2(t)$ (drift coefficient), a function of $X(t)$ and $B(t)$, and $v = 4X(t)B(t)$ (diffusion coefficient) , also a function of $X(t)$ and $B(t)$.

Substituting $X(t) = B^2(t)$ to equation (2.95),

$$\begin{aligned} d(B^4(t)) &= 2(B^2(t) + 2B^2(t)dt + 4(B^2(t))B(t)dB(t), \\ &= 6B^2(t)dt + 4B^3(t)dB(t). \end{aligned} \tag{2.96}$$

We can derive this from chain rule for a function of $B(t)$ as well. Let $g(x) = x^4$ and from Ito formula (equation (2.76)):

$$\begin{aligned} dg &= g'(B(t))dB(t) + \frac{1}{2}g''(t)dt, \\ &= 4B^3(t)dB(t) + \frac{1}{2}4.3.B^2(t)dt. \end{aligned} \tag{2.97}$$

$$d(B^4(t)) = 6B^2(t)dt + 4B^3(t)dB(t). \tag{2.98}$$

This is the same Ito process as in equation (2.98). Let us consider another example which will be useful later in this book. Consider the function $g(x) = \ln x$ and the Ito process

$$dX(t) = \frac{1}{2} X(t) + X(t)dB(t). \tag{2.99}$$

For this Ito process $\mu = \frac{1}{2} X(t)$ and $\sigma = X(t)$.

From Ito formula (equation (2.90)),

$$\begin{aligned} d(\ln X(t)) &= \frac{1}{X(t)} dX(t) + \frac{1}{2}\left(-\frac{1}{X^2(t)}\right)(X^2(t)dt), \\ &= \frac{1}{X(t)}\left(\frac{1}{2} X(t)dt + X(t)dB(t)\right) - \frac{1}{2} dt, \\ &= \frac{1}{2} dt + dB(t) - \frac{1}{2} dt, \\ &= dB(t). \end{aligned} \tag{2.100}$$

By convention, the above stochastic differential is given by the following integral equation:

$$\ln X(t) = \ln X(0) + \int_0^t dB(t), \tag{2.101}$$

$$\ln\left[\frac{X(t)}{X(0)}\right] = B(t), \tag{2.102}$$

$$X(t) = X(0)\, e^{B(t)}.$$

We can show that $X(t) = X(0)e^{B(t)}$ satisfies $dX(t) = \frac{1}{2} X(t)dt + X(t)dB(t)$. In other words $X(t) = X(0)e^{B(t)}$ is a "solution" to the stochastic differential

$$dX(t) = \frac{1}{2} X(t)dt + X(t)dB(t).$$

This idea of having a solution to a stochastic differential is similar to having a solution to differential equations in standard calculus.

2.13.5 Stochastic Product Rule

Suppose $X_1(t)$ and $X_2(t)$ are Ito processes given by the following differentials:

$$dX_1(t) = \mu_1(t)dt + \sigma_1(t)dB(t), \tag{2.103}$$

$$dX_2(t) = \mu_2(t)dt + \sigma_2(t)dB(t). \tag{2.104}$$

Quadratic covariation is given by

$$\begin{aligned} d[X_1, X_2] &= dX_1(t).dX_2(t) \\ &= \mu_1\mu_2(dt)^2 + \mu_1\ \sigma_2 dt.dB(t) + \mu_2\ \sigma_1 dt.dB(t) + \sigma_1\sigma_2(dB(t))^2 \end{aligned}$$

and

$$(dt)^2 = dt.dB(t) = 0.$$

$$\begin{aligned} d[X_1, X_2] &= \sigma_1(t)\sigma_2(t)(dB(t))^2, \\ &= \sigma_1(t)\sigma_2(t)dt. \end{aligned} \tag{2.105}$$

The stochastic product rule is given by,

$$X_1(t)\, X_2(t) - X_1(0)X_2(0) = \int_0^t X_1(s)dX_2(s) + \int_0^t X_2(s)dX_1(s) + [X_1, X_2](t). \tag{2.106}$$

If at least one of X_1 and X_2 is a continuous function with finite variation, then $[X_1, X_2](t) = 0$ and equation (2.106) reduces to the integration by parts formula in the standard calculus.

Stochastic product rule can be expressed in differential form:

$$d(X_1(t)\, X_2(t)) = X_1(t)dX_2(t) + X_2(t)dX_1(t) + \sigma_1(t)\sigma_2(t)dt. \tag{2.107}$$

As an example, consider $Y(t) = t\,B(t)$,

$$Y(t) = X_1(t)X_2(t),$$

where $X_1(t) = t$, a continuous function with finite variation and $\sigma_1 = 0$, and $X_2(t) = B(t)$, Brownian motion with infinite variation and $\sigma_2 = 1$.

From the product rule,

$$d(Y(t)) = tdB(t) + B(t)dt + (0)(1)dt,$$

$$d(tB(t)) = tdB(t) + B(t)dt. \qquad (2.108)$$

This is the same result we obtain if we use the standard product rule. The reason for this is that quadratic covariation of a continuous function and a function with infinite variation is zero as we have discussed previously.
Suppose

$$dX_1(t) = tdB(t) + B(t)dt, \text{ and}$$
$$dX_2(t) = \frac{1}{2}X_2(t)dt + X_2(t)dB(t),$$

where

$$\mu_1(t) = B(t);\ \sigma_1(t) = t;\ \sigma_2(t) = X_2(t);\ \text{and}\ \mu_2(t) = \frac{1}{2}X_2\ .$$

From the product rule,

$$\begin{aligned} d(X_1(t)X_2(t)) &= X_1(t)dX_2(t) + X_2(t)dX_1(t) + \sigma_1\sigma_2 dt, \\ &= X_1(t)dX_2(t) + X_2(t)dX_1(t) + tX_2(t)dt. \end{aligned} \qquad (2.109)$$

By substitution,

$$\begin{aligned} d(X_1(t)X_2(t)) &= X_1(t)(\frac{1}{2}X_2dt + X_2(t)dB(t)) \\ &\quad + X_2(t)(tdB(t) + B(t)dt) + tX_2(t)dt, \\ &= \left(\frac{1}{2}X_1(t)X_2(t) + X_2(t)B(t) + tX_2(t)\right)dt \\ &\quad + (X_1(t)X_2(t) + tX_2(t))dB(t). \end{aligned} \qquad (2.110)$$

This is again an Ito process.

$$\begin{aligned} d(X_1(t)X_2(t)) &= \left(\frac{1}{2}X_1(t)X_2(t) + tX_2(t) + X_2(t)B(t)\right)dt \\ &\quad + \left(X_1(t)+t\right)X_2(t)dB(t), \\ &= X_2(t)\left(\frac{1}{2}X_1(t)+t+B(t)\right)dt \\ &\quad + X_2(t)\left(X_1(t)+t\right)dB(t)). \end{aligned} \tag{2.111}$$

As an integral equation,

$$X_1(t)X_2(t) - X_1(0)X_2(0) = \int_0^t X_2(t)(\frac{1}{2}X_1(t)+t+B(t))dt + \int_0^t X_2(t)(X_1(t)+t)dB(t).$$

2.13.6 Ito Formula for Functions of Two Variables

If $g(x_1, x_2)$ is a continuous and twice differentiable function of x_1 and x_2 and Ito processes are of the forms

$$dX_1(t) = \mu_1 dt + \sigma_1\, dB(t) \tag{2.112}$$

and

$$dX_2(t) = \mu_2 dt + \sigma_2 dB(t)\ . \tag{2.113}$$

Then $g(X_1(t), X_2(t))$ is also an Ito process and given by the following differential form:

$$\begin{aligned} dg(X_1(t), X_2(t)) &= \frac{\partial g(X_1)}{\partial x_1}dX_1(t) + \frac{\partial g(X_2)}{\partial x_2}dX_2(t) + \frac{1}{2}\frac{\partial^2 g(X_1)}{\partial^2 x_1}(dX_1(t))^2 \\ &\quad + \frac{1}{2}\frac{\partial^2 g(X_2)}{\partial^2 x_2}(dX_2(t))^2 + \frac{\partial^2 g(X_1 X_2)}{\partial x_1 x_2}dX_1(t).dX_2(t). \end{aligned} \tag{2.114}$$

Using quadratic variation and covariation of Ito processes,

$$(dX_1(t))^2 = dX_1(t)\,.\,dX_1(t) = \sigma_1^2 dt\,,$$

$$(dX_2(t))^2 = dX_2(t)\,.\,dX_2(t) = \sigma_2^2 dt\,, \quad \text{and} \tag{2.115}$$

$$dX_1(t)\,.\,dX_2(t) = \sigma_1\sigma_2 dt.$$

These can be considered as a generalization of the rules on differentials given by equation (2.74). We use this generalized Ito formula for a function of two Ito processes in the following example.

We will express the stochastic process $X(t) = 2 + t + e^{B(t)}$ as an Ito process having the standard form, $dX(t) = \mu dt + \sigma dB(t)$.
We can consider

$$X(t) = g(t, B(t)) = 2 + t + e^{B(t)}. \tag{2.116}$$

Therefore, $g(t, y) = 2 + t + e^{y}$,

where $X_1(t) = t$,
$X_2(t) = y = B(t)$.

These equations give, $dX_1 = dt$ and $dX_2 = dB(t)$,
where $\mu_1 = 1$; $\sigma_1 = 0$; $\mu_2 = 0$; and $\sigma_2 = 1$.

Using equation (2.114),

$$\begin{aligned} dg &= (1)dt + e^{B(t)}dB(t) + \frac{1}{2}(0)(dB(t))^2 + \frac{1}{2}e^{B(t)}(dB(t))^2 \\ &\quad + (0)dt.dB(t), \\ &= dt + e^{B(t)}dB(t) + \frac{1}{2}e^{B(t)}dt. \end{aligned} \tag{2.117}$$

Using $dB(t))^2 = dt$,

$$\begin{aligned} dg &= dX(t), \\ &= \left(1 + \frac{1}{2}e^{B(t)}\right)dt + e^{B(t)}dB(t). \end{aligned} \tag{2.118}$$

From a previous example,

$$d(e^{B(t)}) = e^{B(t)}dB(t) + \frac{1}{2}e^{B(t)}dt. \quad (2.119)$$

Therefore

$$\begin{aligned} dX(t) &= dt + (\frac{1}{2}e^{B(t)}dt + e^{B(t)}dB(t)), \\ &= dt + d(e^{B(t)}). \end{aligned} \quad (2.120)$$

From the integral notation,

$$X(t) = X(0) + \int_0^t dt + \int_0^t d(e^{B(t)}), \quad (2.121)$$

$$X(t) = (0) + t + e^{B(t)} - 1,$$

$$X(t) = (X(0) - 1) + t + e^{B(t)}. \quad (2.122)$$

Comparing with

$$\begin{aligned} &X(t) = 2 + t + e^{B(t)}, \\ &X(0) - 1 = 2, \\ &X(0) = 3. \end{aligned}$$

$X(t)$ = constant + t + $e^{B(t)}$ can be considered as a solution process to the stochastic differential,

$$dX(t) = (1 + \frac{1}{2}e^{B(t)})dt + e^{B(t)}dB(t).$$

As we can see in the above solution, the solution process contains the characteristics of both the drift and diffusion phenomena. In this case, diffusion phenomenon dominates as t increases because of the expected value of the exponential of Brownian motion increases at a faster rate in general. If we examine the drift term of the stochastic differential above, we see that the drift term is also affected by the Brownian motion, so the final solution is always a result of complex interactions between the drift term and the diffusion term.

2.14 Stochastic Population Dynamics

We will now move back to discuss the population dynamics example equipped with the knowledge of Ito process and formula:

$$\frac{dx(t)}{dt} = \alpha(t)x(t)\,. \tag{2.123}$$

If $\alpha(t) = r(t) + \sigma(t)W_t$,

where

W_t = white noise,

then

$$dX(t) = (r(t)dt + \sigma(t)dB(t))\,.\,X(t)\,,$$

and Brownian motion increments $dB(t) = W_t dt$.

Therefore ,

$$dX(t) = \left(r(t)dt + \sigma(t)W_t dB(t)\right)X(t)\,,$$
$$dX(t) = r(t)X(t)dt + \sigma(t)X(t)d(t)\,. \tag{2.124}$$

As seen from the above equation (2.124), $X(t)$ is an Ito process.

Consider the case with $r(t) = r$, a constant and $\sigma(t) = \sigma$, a constant then the process $X_1(t)$ can be written in the differential form:

$$dX(t) = rX(t)dt + \sigma X(t)dB(t)\,. \tag{2.125}$$

Assume $g(x)=\ln x$, and using the Ito formula,

$$dg(x(t)) = \frac{\partial(\ln x)}{\partial x}dX(t) + \frac{1}{2}\frac{\partial^2 g}{\partial x^2}(dX(t))^2\,,$$

$$d(\ln(X(t)) = \frac{dX(t)}{X(t)} + \frac{1}{2}\left(\frac{-1}{X(t)^2}\right)(\sigma^2 X^2(t)(dB(t))^2),$$
$$= \frac{dX(t)}{X(t)} - \frac{\sigma^2}{2}dt = \frac{1}{X(t)}(rX(t)dt + \sigma X(t)dB(t)) - \frac{\sigma^2}{2}dt.$$

$$d(\ln(X(t)) = rdt + \sigma dB(t) - \frac{\sigma^2}{2}dt,$$
$$= \left(r - \frac{\sigma^2}{2}\right)dt + \sigma dB(t).$$

Converting back to the integral form,

$$\ln(X(t)) = \ln(X(0)) + \int_0^t \left(r - \frac{\sigma^2}{2}\right)dt + \int_0^t \sigma dB(t),$$

$$\ln\left(\frac{X(t)}{X(0)}\right) = \left(r - \frac{\sigma^2}{2}\right)t + \sigma B(t),$$

$$X(t) = X(0)\exp(\sigma B(t))\exp\left((r - \frac{\sigma^2}{2})t\right). \tag{2.126}$$

$X(t)$ process, therefore, satisfies the Ito process,

$$dX(t) = rX(t)dt + \sigma X(t)dB(t),$$

and equation (2.126) can be considered as a solution to the stochastic differential equation. As discussed earlier, this solution significantly different from its deterministic counter part.

In this chapter we have revised the essential results from stochastic calculus and presented the results which could be useful in developing models and solving stochastic differential equations. While analytical expressions are quite helpful to understand stochastic processes, computer simulation provides us with an intuitive "feel" for the simulated phenomena. Sometimes it is revealing to simulate a number of realizations of a process and visualize them on computers to understand the behavior of the process. We will devote the next chapter to computer simulation of Brownian motions, Ito integrals, Ito processes and the solutions of a limited number of stochastic differential equations.

Chapter 3

Computer Simulation of Brownian Motion and Ito Processes

3.1 Introduction

In Chapter 2, we have introduced Brownian motion (the Wiener process) as a stationary, continuous stochastic process with independent increments. This process is a unique one to model the irregular noise such as Gaussian white noise in systems, and once such a process is incorporated in differential equations, the process of obtaining solutions involve stochastic calculus. Only a limited number of stochastic differential equations have analytical solutions and some of these equations are given by Kloeden and Platen (1992). In many instances we have to resort to numerical methods. The objective of this chapter is to illustrate the behavior of the Wiener process and Ito processes through computer simulations so that reader can appreciate the variable nature of individual realizations. The routines are written in Mathematica® (1999). Some of these routines may be useful in constructing numerical solutions of stochastic differential equations later in this book .

3.2 A Standard Wiener Process Simulation

For the numerical implementation, it is most convenient to use the variance specification of the Wiener process *B(t)* in the form of equation (2.21). The time span of the simulation, [0,1] is discretized into small equal time increments *delt* , and the corresponding independent Wiener increments selected randomly from a normal distribution with zero mean and standard deviation equal to $\sqrt{delt}$. This is implemented by the following simple Mathematica program:

```
<< Statistics`ContinuousDistributions`

wienerProcess[delt_, Ni_] := Module[{std, ii = 1, kk = 1},
(*Ni---number of intervals; delt--- interval *)
std = N[Sqrt[delt], 32];
incrementList = Table[N[Random[NormalDistribution[0.0, std]], 32], {Ni}];
incrementList = PrependTo[incrementList, 0.0];
standardWiener = N[Apply[Plus, Table[Table[incrementList[[ii]],
                {ii, 1, kk}], {kk, 1, Ni + 1}], {1}], 32];];
wienerProcess[0.001, 1000] // Timing
ListPlot[standardWiener, PlotJoined -> True, PlotRange -> All] // Timing
```

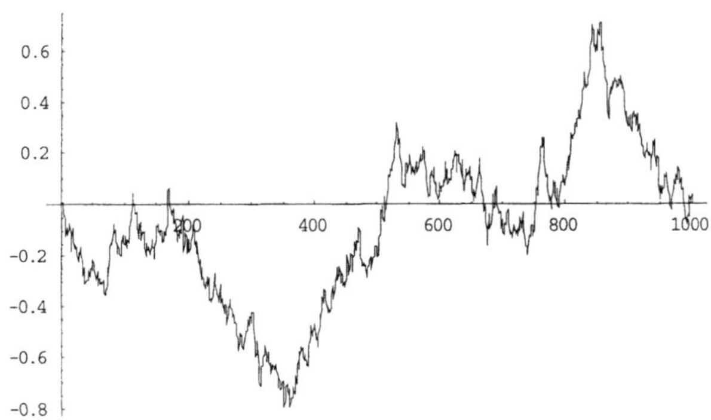

Figure 3.1 Mathematica program for the standard Wiener process and a sample realization.

In the example given above we have used the definition of Ito integral by evaluating the increment during the interval $[i - 1, i]$ at $(i - 1)$ and computing the Wiener process by adding the increment to the value of Brownian motion at $[i - 1]$. In this example the time interval [0, 1] is divided into very small equal divisions, and the graph is shown in terms of the number of time intervals instead of time t. We have generated 1000 Brownian motion increments that are Gaussian random variables, and Figure 3.2 shows these increments as a single stochastic process. Since Gaussian white noise is the derivative of Brownian motion and as the time interval is a constant, Figure 3. 2 depicts a realization of a white noise process.

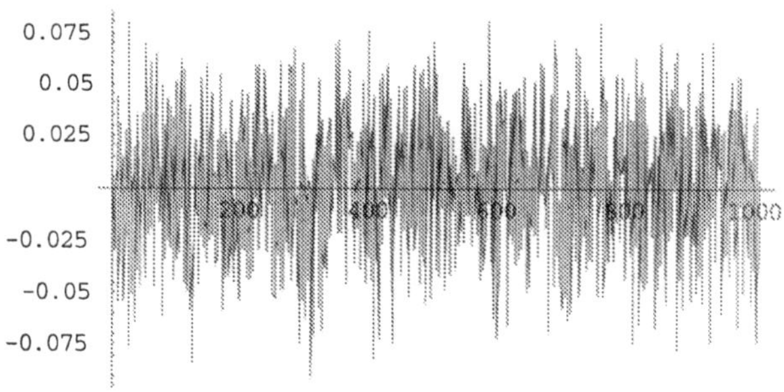

Figure 3.2 A realization of Brownian increments.

In Figure 3.1, the realization shown tend to come back to original position, but Figure 3.3 shows a significant diversion from the origin.

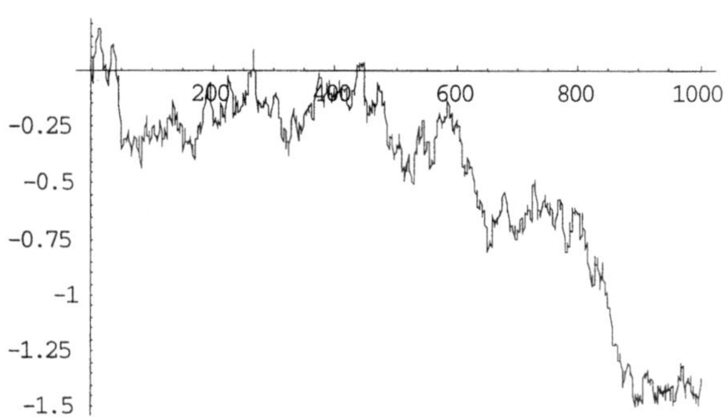

Figure 3.3 Another realization of the standard Brownian motion.

These two realizations have quite different directions of movement, even though expected value of Brownian motion at a given time is zero. To investigate the behavior further we have produced 10 realizations in Figure 3.4.

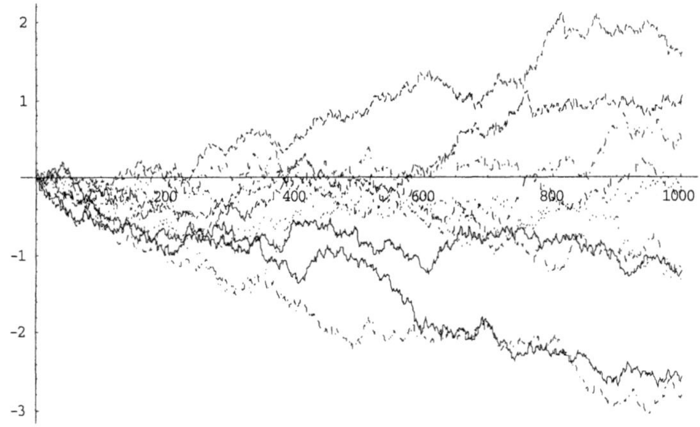

Figure 3.4 Ten different realizations of Brownian motion.

It is seen that the motion is very irregular, and the only discernible pattern is that as time progresses, the position tends to wander away from the starting position at the origin. In other words, if the statistical variance over realizations for a fixed time is evaluated, this increases gradually – a property referred to as time varying variance.

The use of the process in a modeling situation to represent the noise in the system should be carefully thought through. If the noise can be represented as white noise, then Brownian motion enters into the equation because of the relationship between the white noise and Brownian motion as a Gaussian white noise process can be approximated as the derivative of Brownian motion.

3.3 Simulation of Ito Integral and Ito Processes

It is important to realize that the Ito integral is a stochastic process dependent on the Wiener process. This is analogous to integration in standard calculus because an indefinite integral is a function of the independent, deterministic variable. Given the Brownian motion realization depicted in Figure 5, we will compute the Ito integral of Brownian motion:

$$\int_0^t B(t,\omega)dB .$$

As we have previously seen, this integral can be evaluated by using the following stochastic relationship converging in probability:

$$\int_0^t B(s,\omega)dB(s,\omega) = \frac{1}{2}B^2(t,\omega) - \frac{1}{2}t . \qquad (3.1)$$

We have computed this Ito integral using the following Mathematica statement taking the time interval as 0.001 which was the value used for the generation of the Brownian realization in Figure 3.5:

```
itoIntegralBdB = Table[0.5 (standardWiener[[i]])^2 + 0.5 0.001 (i - 1), {i, 1, 1000}];
```

The corresponding realization is given in Figure 3.6.

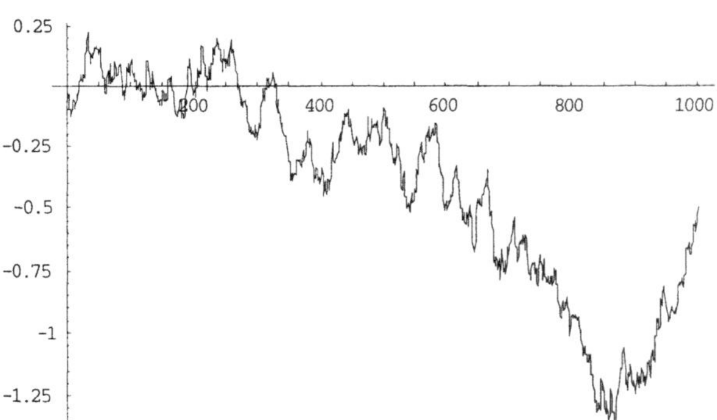

Figure 3.5 The realization of the Wiener process used in the calculation of the Ito Integral depicted in Figure 3.6.

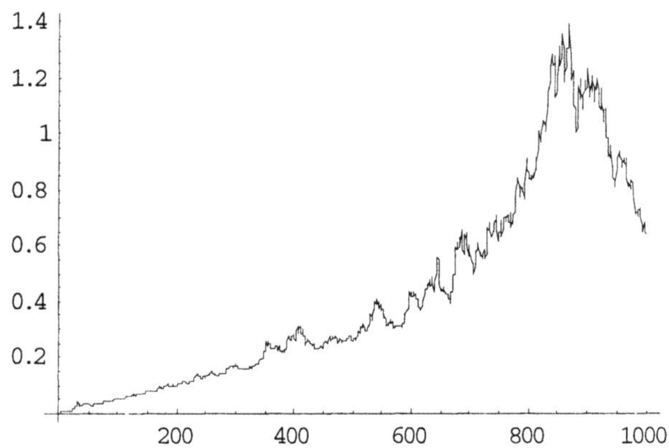

Figure 3.6 A realization of $\int_0^t B(t,\omega)dB$.

As seen in Figures 3.5 and 3.6, the realization of the Ito integral depends strongly on the square of Brownian motion as time increases and this tendency weakens as time advances beyond 1.

Next we will compute another realization of Brownian motion (Figure 3.7) and corresponding Ito integral $\int_0^t B(t,\omega)dB$ (Figure 3.8).

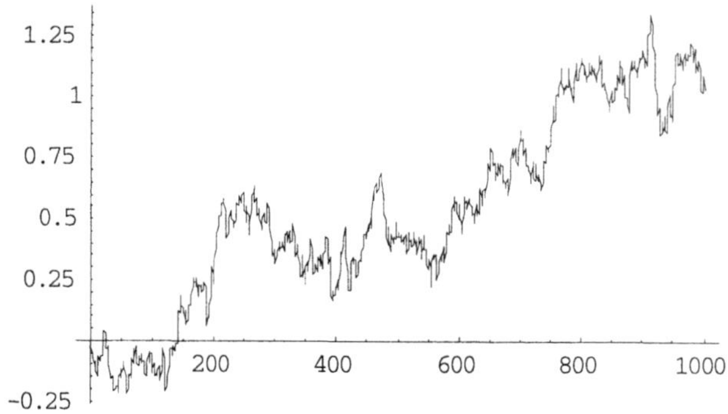

Figure 3.7 Another realization of Brownian motion.

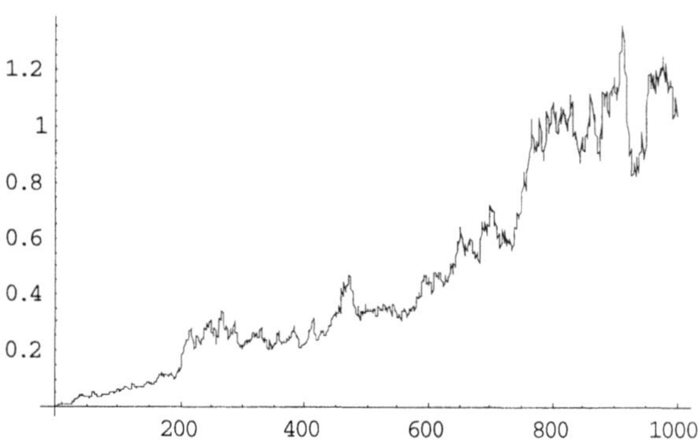

Figure 3.8 Ito integral $\int_0^t B(t,\omega)dB$ corresponding to the Brownian realization in Figure 3.7.

Let us consider the following Ito process which we have derived in Chapter 2. In differential notation,

$$d(B^4(t)) = 6B^2(t)dt + 4B^3(t)dB(t),$$

which means,

$$B^4(t) = B^4(0) + \int_0^t 6B^2(t)dt + \int_0^t 4B^3(t)dB(t), \text{ and}$$

$$B^4(t) = \int_0^t 6B^2(t)dt + \int_0^t 4B^3(t)dB(t) \ . \tag{3.2}$$

This Ito process has a drift term as well as a diffusion term, and the process can be evaluated by using the following Mathematica code given that we have evaluated a new standard Wiener realization:

```
itoProcessB4 = Table[ Sum[ (6 (standardWiener[[j]])^2 0.001 +
        4 (standardWiener[[j]])^3 incrementList[[j]]), {j, 1, i}], {i, 1, 999}];
```

The Ito process given in equation (3.2) is simulated in Figure 3.10 for the Wiener realization depicted in Figure 3.9.

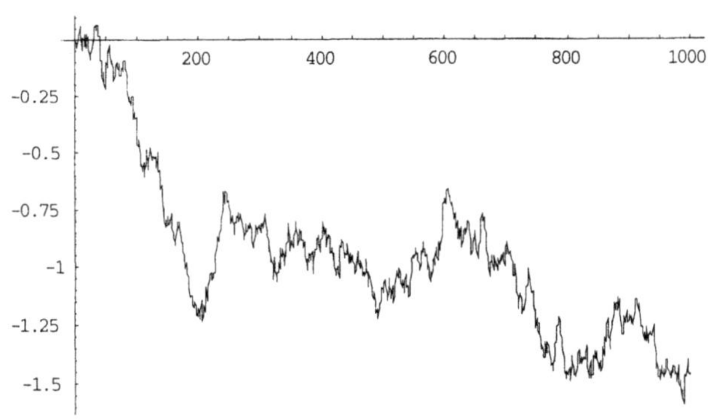

Figure 3.9 Wiener realization used in evaluating the Ito process $B^4(t)$ as seen in Figure 3.10.

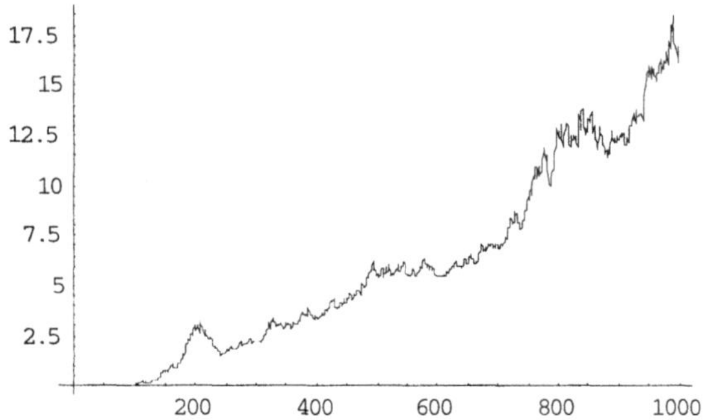

Figure 3.10 Ito process $B^4(t)=\int_0^t 6B^2(t)dt+\int_0^t 4B^3(t)dB(t)$.

Even for a decreasing and erratic Brownian motion, the Ito process $\left\{\int_0^t 6B^2(t)dt+\int_0^t 4B^3(t)dB(t)\right\}$ in general has a smoother realization which has an overall growth in positive direction. The effect of Ito integration tends to smother the erratic behavior of Brownian motion. We have evaluated the above Ito process for 3 different realizations of the standard Wiener process, and they are shown in Figure 3.11.

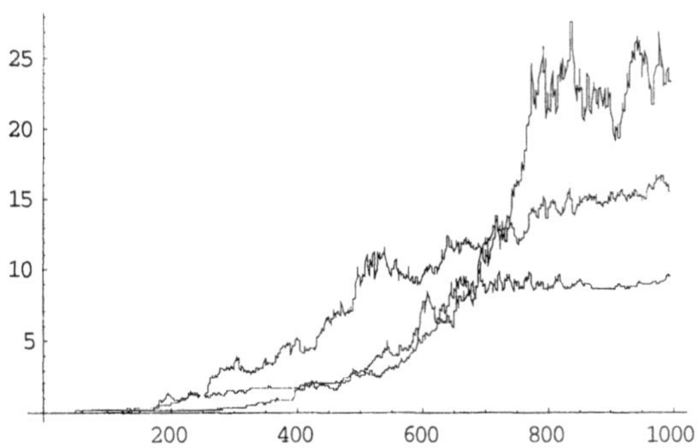

Figure 3.11 Three realizations of $\left\{\int_0^t 6B^2(t)dt+\int_0^t 4B^3(t)dB(t)\right\}$.

As seen in Figure 3.11, individual realizations of the Ito process $\left\{\int_0^t 6B^2(t)dt + \int_0^t 4B^3(t)dB(t)\right\}$ are distinct from each other and therefore shows the complexity in stochastic integration as opposed to integration in the standard calculus, and this illustrates that it is important to simulate the stochastic processes in applications to better understand the variability of observations. Among a large number of realizations, one may observe extreme events which can not be mathematically obtained. This leads us to discuss a specific stochastic model and we will go back to our stochastic population dynamic model.

3.4 Simulation of Stochastic Population Growth

We consider equation (2.5), which is the solution to the population growth model with a variable coefficient. The Mathematica® code for the solution of equation (2.5) is given below for $r = 1.5$ and $\sigma = 1.0$. Note that for these particular values, the coefficients of t and B_t in the exponents both reduce to a value of 1. Figure 3.13 shows a sample of realizations (or sample paths) of the solution and the horizontal axis gives the number of time intervals.

```
(* initial value of the population=1.0 *)

Do[
        Module[{t=0,i=1,delt= 2^-9}, B[0]=0.0;
                    s=delt ^0.5;
        x[0]=1.0;       xd[0]=1.0;
        While[t< 1.0,
                B[i]=B[i-1]+ Random[NormalDistribution[0.0,s]];
                x[i]=x[0] Exp[t + B[i-1]];
                xd[i]=xd[0] Exp[t];
                t=t+delt;   i++
        ];
        Print[i-1];
        ListPlot[Table[x[j],{j,1,i-1}],PlotJoined->True]
        ],
{n,1,10}]
```

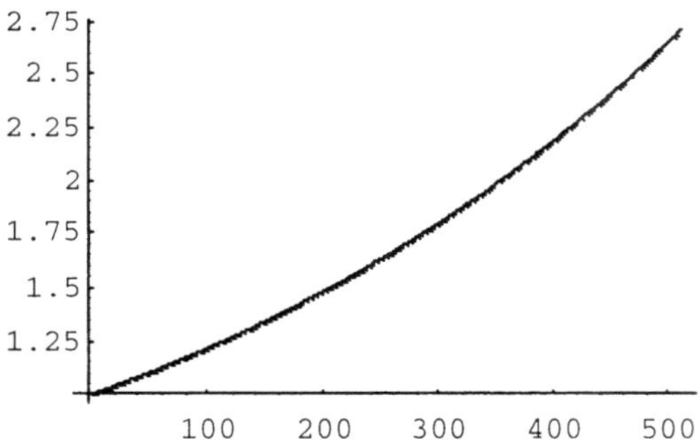

Figure 3.12 Deterministic solution of population growth equation.

Figures 3.12 and 3.13 show the deterministic solution and a sample of different realizations of the stochastic solution. The most striking aspect is how little of the behavior of the deterministic exponential growth curve remains recognizable in the stochastic realizations. In fact, it would be difficult to distinguish by inspection between the realizations of the simple Brownian motion and those of exponential growth in Figure 3.13.

To an observer, any one of these realizations in Figure 3.13 can be seen as an outcome of the process. A limited number of samples will obscure the fact that these realizations result from a mechanistic relationship, but with noisy coefficients with irregular behavior. This also shows that the variability in parameters can significantly change the outcome of the process.

These observations apply, of course, to the cases where the amplitude of the noise term is comparable to the growth rate. Figure 3.14 shows realizations with $\sigma = 0.5$ and $r = 1.5$ and comparing with Figure 3.13, the underlying exponential growth can now be recognized.

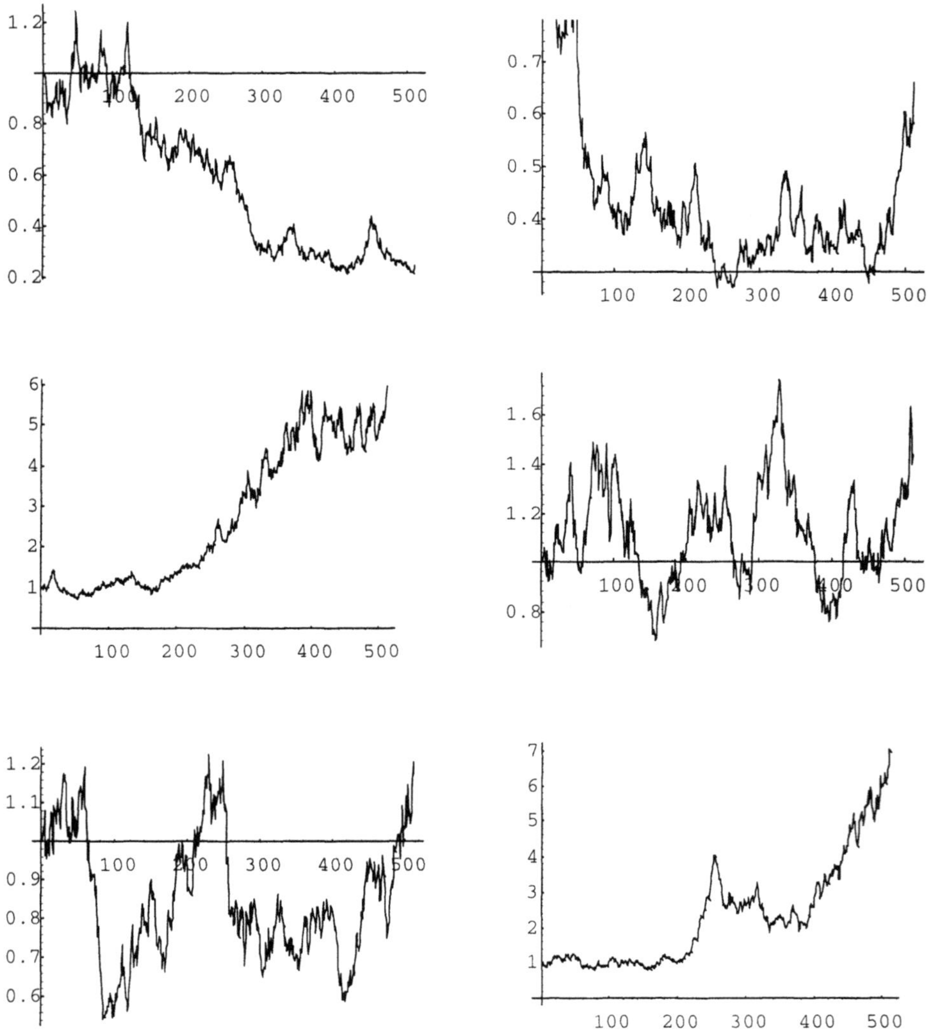

Figure 3.13 A sample of different realizations of equation (2.5).

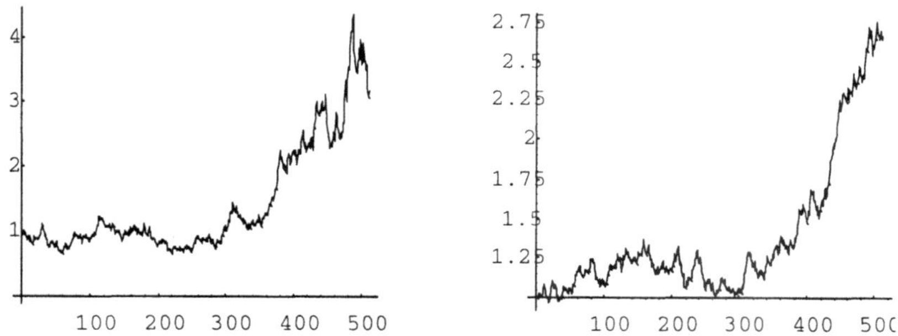

Figure 3.14 Realizations of equation (2.5) with $\sigma = 0.5$ and $r = 1.5$.

If the deterministic solution is not known, the statistical moments such as the mean and the variance could be calculated from a sufficiently large sample of realizations at a given time, *t*, to discuss the behavior of the system. It seems reasonable to expect that such mean values will approximate the behavior of the deterministic solution and allow determination of the growth rate, provided that σ can also be determined. However, the variance depends on both σ and the time at which the sample is taken. Clearly, the extraction of model parameters from a limited set of individual realizations is not straightforward.

Moreover, as will be shown in Chapter 5, the intuitive expectation about the representativeness of a finite set of samples is not fully justified. While the mean over all realizations indeed follows the deterministic behavior, it turns out that the majority of realizations do not reach the deterministic population value at a given time. Hence a finite sample is likely to underestimate the underlying growth rate. To fully explore this phenomenon requires considerable additional theoretical background and that is the subject of Chapter 5.

Leaving these issues aside, we note that Figures 3.13 and 3.14 show the effects of σ on the variance, and this provides us with a way of constructing confidence intervals for the results of experiments. Confidence intervals along with the moments should be used to validate the model given by equation (2.1) with "field" data from experiments. Therefore, when models are constructed using stochastic differential equations with Brownian motion to represent the noise in the system, they can be used to conduct computer experiments to understand and predict the behavior of the systems under study. The results of the experiments should be analyzed and interpreted using appropriate statistical methods. For simple models such as the one given in equation (2.5), solutions can be found and the need for extensive statistical analysis of results is not necessary. But for complex systems, incorporation of noise can result in mathematically unpredictable behavior; therefore, computer experimentation with the system models is the only way of examining the randomness affecting different parameters in the model. The only validation that can be done is to compare results from computer experiments with data from the actual system. (See Brown and Kulasiri (1996) for a discussion of validation of complex, stochastic, biological systems.)

Chapter 4

Solving Stochastic Differential Equations

4.1 Introduction

In Chapter 2, we discussed the elementary concepts in stochastic calculus and showed in a limited number of situations how it differs from the standard calculus. We have defined Ito integrals and introduced Ito processes along with some of the tools that could be useful in working with stochastic calculus. In this chapter, we intend to review stochastic differential equations (SDEs) briefly and the ways of solving them analytically. The main aim of this chapter is to present a very limited number of solution methods which are useful within the context of the scope of this book but more importantly to encourage reader to pursue this subject using more rigorous treatments available.

4.2 General Form of Stochastic Differential Equations

Let us consider an ordinary differential equation which relates the derivative of the dependent variable ($y(t)$) to the independent variable (t) through a function, $\phi(y(t),t)$, with the initial condition $y(0) = y_0$:

$$\frac{dy}{dt} = \phi(y,t), \tag{4.1}$$

and

$$dy = \phi(y,t)dt .$$

In many natural systems, this rate of change can be influenced by random noise caused by a combination of factors, which could be difficult to model. As a model of this random fluctuations, white noise ($\xi(t)$) is a suitable candidate. Therefore we can write, in general, the increments of the noise process as $\sigma(y,t)\,\xi(t)$ where σ is an amplitude function modifying the white noise.

Hence,

$$\frac{dy}{dt} = \phi(y,t) + \sigma(y,t)\,\xi(t). \tag{4.2}$$

As we have see from Chapter 2,

$$\sigma(y,t)\,\xi(t) = \sigma(y,t)\frac{dB}{dt} \tag{4.3}$$

where, $B(t)$ = Brownian motion.

Therefore,

$$\frac{dy}{dt} = \phi(y,t) + \sigma(y,t)\frac{dB}{dt}, \tag{4.4}$$

$$dy = \phi(y,t)d\phi + \sigma(y,t)dB. \tag{4.5}$$

In general, $\phi(y,t)$ and $\sigma(y,t)$, could be stochastic processes. This equation is called a stochastic differential equation (SDE) driven by Brownian motion. Once Brownian motion enters into equation (4.3), y becomes a stochastic process, $Y(t,\omega)$, and in the differential notation SDE is written as

$$dY(t) = \phi(Y(t),t)dt + \sigma(Y(t),t)\,dB(t). \tag{4.6}$$

This actually means,

$$Y(t) = Y(0) + \int_0^t \phi(Y(t),t)\,dt + \int_0^t \sigma(Y(t),t)\,dB(t). \tag{4.7}$$

If we can find a function of Brownian motion in the form of an Ito process that satisfies the above integral equation (4.7), we call that function a strong solution of SDE.

Strong solutions do not depend on individual realizations of Brownian motion. In other words, all possible realizations of an Ito process, which is a strong solution of a SDE, satisfy the SDE under consideration. Not all the SDEs have strong solutions. The other class of solutions are called weak solutions where solution to each individual realization is different from each other. In this chapter we will focus only on strong solutions. In many situations, finding analytical solutions to SDEs is impossible and therefore we will review a minimum number of SDEs and their solutions so that reader can embark on learning this challenging area of applied mathematics.

4.3 A Useful Result

If $X(t)$ is a stochastic process and another stochastic process $Y(t)$ is related to $X(t)$ through the stochastic differential,

$$dY(t) = Y(t)\, dX(t)\,, \tag{4.8}$$

with $Y(0) = 1$.

Thus $Y(t)$ is called the stochastic exponential of $X(t)$. If $X(t)$ is a stochastic process of finite variation, then the solution to equation (4.8) is,

$$Y(t) = e^{X(t)}\,, \tag{4.9}$$

and, for any process $X(t)$,

$Y(t) = e^{\xi(t)}$ satisfies the stochastic differential given above when

$$\xi(t) = X(t) - X(0) - \frac{1}{2}[X, X](t)\,. \tag{4.10}$$

$[X, X](t)$ is quadratic variation of $X(t)$ and for a continuous function with finite variation $[X,X](t) = 0$.

For example, consider the following stochastic differential equation in differential form,

$$dX(t) = X(t)\,dB(t). \tag{4.11}$$

This SDE does not have a drift term and the diffusion term is an Ito integral. We know, $[B, B](t) = t$.

Therefore from the above result,

$$\begin{aligned}\xi(t) &= B(t) - B(0) - \frac{1}{2}t,\\ &= B(t) - \frac{1}{2}(t).\end{aligned} \tag{4.12}$$

Then the solution to the SDE is

$$X(t) = e^{B(t)-\frac{1}{2}t}. \tag{4.13}$$

Now let us consider a similar SDE with a drift term:

$$dX(t) = \alpha\, X(t)\,dt + \beta\, X(t)\,dB(t), \tag{4.14}$$

where α and β are constants.

Dividing it by $X(t)$,

$$\frac{dX(t)}{X(t)} = \alpha\; dt + \beta\; dB(t). \tag{4.15}$$

This differential represents,

$$\begin{aligned}\int_0^t \frac{dX(t)}{X(t)} &= \int_0^t \alpha\, dt + \int_0^t \beta\, dB(t),\\ &= \alpha\, t + \beta(B(t) - B(0)),\\ &= \alpha\, t + \beta\, B(t).\end{aligned} \tag{4.16}$$

The second term on the right hand side comes from Ito integration.

Now let us assume

$$\phi(t) = \alpha t + \beta B(t).$$

Then the SDE becomes,

$$\int_0^t \frac{dX(t)}{X(t)} = \phi(t). \tag{4.17}$$

$X(t)$ is a stochastic exponential of $\phi(t)$ with corresponding $\xi(t)$:

$$\xi(t) = \phi(t) - \phi(0) - \frac{1}{2}[\phi, \phi](t). \tag{4.18}$$

$$\begin{aligned}[\phi,\phi](t) &= [(\alpha t + \beta B(t)), (\alpha t + \beta B(t))](t), \\ &= [\alpha t, \alpha t](t) + 2\alpha \beta [t, B(t)](t) + \beta^2 [B, B](t), \\ &= 0 + 0 + \beta^2 t.\end{aligned} \tag{4.19}$$

Therefore,

$$\xi(t) = \alpha t + \beta B(t) - 0 - \frac{1}{2}\beta^2 t.$$

The solution to the SDE is

$$X(t) = \exp((\alpha - \frac{1}{2}\beta^2)t + \beta B(t)). \tag{4.20}$$

Let us examine whether the stochastic process

$$X(t) = \exp((\alpha - \frac{1}{2}\beta^2)t + \beta B(t))$$

is a strong solution to the differential equation

$$dX(t) = \alpha X(t)dt + \beta X(t)\, dB(t). \tag{4.21}$$

We will define a function,

$$f(x,t) = \exp((\alpha - \frac{1}{2}\beta^2)t + \beta\, x).$$

Then

$$\begin{aligned} X(t) &= f(B(t),t), \\ &= \exp((\alpha - \frac{1}{2}\beta^2)t + \beta\, B(t)). \end{aligned} \tag{4.22}$$

We need to apply Ito formula for the two Ito processes $X_1(t)$ and $X_2(t)$ (equation (2.114)).

$X_1(t) = B(t)$; $X_2(t) = t$ (a continuous function with finite variation);

$dX_1.dX_2(t) = d[X_1, X_2] = 0$; $(dX_1)^2 = dt$; $(dX_2)^2 = 0$.

Differentiating the function f with respect to x,

$$\frac{\partial f}{\partial x} = \beta \exp((\alpha - \frac{1}{2}\beta^2)t + \beta x),$$

and differentiating again w.r.t. x,

$$\frac{\partial^2 f}{\partial x^2} = \beta^2 \exp((\alpha - \frac{1}{2}\beta^2)t + \beta x).$$

Differentiating f with respect to t,

$$\frac{\partial f}{\partial t} = (\alpha - \frac{1}{2}\beta^2)\exp((\alpha - \frac{1}{2}\beta^2)t + \beta x).$$

From Ito formula (equation 2.114),

$$\begin{aligned} d(f(X_1, X_2)) &= \frac{\partial f}{\partial x} dB(t) + \frac{\partial f}{\partial t} dt + \frac{1}{2}\frac{\partial^2 f}{\partial x^2} dt + \frac{1}{2}\frac{\partial^2 f}{\partial t^2}(0) + \frac{1}{2}\frac{\partial^2 f}{\partial x\, \partial t}(0), \\ &= \beta \exp((\alpha - \frac{1}{2}\beta^2)t + \beta\, B(t)) + (\alpha - \frac{1}{2}\beta^2)\exp(\alpha - \frac{1}{2}\beta^2)dt \\ &\quad + \frac{1}{2}\beta^2 \exp(\alpha - \frac{1}{2}\beta^2)dt. \end{aligned} \tag{4.23}$$

$$
\begin{aligned}
d(X(t)) &= d(f(B(t),t)), \\
&= \alpha \exp((\alpha - \frac{1}{2}\beta^2)t + \beta\, B(t))dt \\
&\quad + \beta \exp((\alpha - \frac{1}{2}\beta^2)t + \beta\, B(t))dB(t), \\
&= \alpha X(t)dt + \beta X(t)dB(t).
\end{aligned}
\tag{4.24}
$$

This proves that $X(t) = f(B(t),t)$ is a strong solution of the SDE given by equation (4.22).

We can see that if we can find a function $f(x,t)$, and for a given Brownian motion $B(t)$, $X(t) = f(B(t),t)$ is a solution to the SDE of the form

$$dX(t) = \mu(X(t),t)dt + \sigma(X(t),t)\, dB(t).$$

$X(t)$ should also satisfy,

$$X(t) = X(0) + \int_0^t \mu(X(s),s) + \int_0^t \sigma\, dB(s), \tag{4.25}$$

provided that $\int_o^t \mu\, ds$ and $\int_o^t \sigma dB(s)$ exist.

4.4 Solution to the General Linear SDE

Solution to the general linear SDE of the form,

$$dX(t) = (\alpha(t) + \beta(t)X(t))dt + (\gamma(t) + \delta(t)X(t))\, dB(t), \tag{4.26}$$

where α, β, γ and δ are given adapted processes and continuous functions of t, can be quite useful in applications.

The solution can be expresses as a product of two Ito processes (Klebaner, 1998)

$$X(t) = u(t)\, v(t), \tag{4.27}$$

where

$$du(t) = \beta\, u(t)dt + \delta u(t)\, dB(t), \tag{4.28}$$

and

$$dv(t) = a\, dt + b\, dB(t). \tag{4.29}$$

$u(t)$ can be solved by using a stochastic exponential as shown above and once we have a solution, we can obtain $a(t)$, $b(t)$ by solving the following two equations:

$$b(t)\, u(t) = \gamma(t), \text{ and} \tag{4.30}$$

$$a(t)\, u(t) = \alpha(t) - \delta(t)\, \gamma(t). \tag{4.31}$$

Then the solution to the general linear SDE is given by (Klebaner, 1998):

$$X(t) = u(t)\left(X(0) + \int_0^t \frac{\alpha(s) - \delta(s)\, \gamma(s)}{u(s)} ds + \int_0^t \frac{\gamma(s)}{u(s)} dB(s) \right). \tag{4.32}$$

As an example let us solve the following linear SDE:

$$dx(t) = a\, X(t)\, dt + dB(t), \tag{4.33}$$

where a is a constant.

Here $\beta(t)=a$, $\gamma(t)=1$, $\alpha(t)=0$, and $\delta(t)=0$.

Using the general solution with

$$du(t)=a\,u(t)\,dt+(0)\,dB(t),$$
$$=a(t)\,u(t)\,dt.$$

From stochastic exponential, $u(t)=\exp(at)$.

Therefore,

$$X(t)=\exp(at)(X(o)+\int_{o}^{t}\exp(-a\,s)\;dB(s)). \tag{4.35}$$

This is a strong solution of the SDE given by equation (4.35).

The integral in the solution given above is an Ito integral and should be calculated according Ito integration. For nonlinear stochastic differential equations, some times appropriate substitutions can be found to reduce them to linear ones. In some situations, we can simplify a nonlinear SDE to a linear one to study the approximate behavior of the system. In these situations analytical results provide insightful information about the system behavior.

Chapter 5

Potential Theory Approach to SDEs

5.1 Introduction

The examples in the previous chapter make it clear that the behavior of individual realizations of a stochastic process are often too variable to give a satisfactory account of the behavior of the physical system that the process is designed to model. Instead, one has to simulate many realizations, and draw conclusions from an appropriate statistical analysis of the results.

This conclusion is reinforced if one considers in more detail the physical interpretation of an individual realization. For example, in the porous flow problem, a single realization of the solution to the SDE that models liquid flow, might give the path followed by a single fluid element through the porous medium starting from a given initial position at a given initial time, such as that illustrated in Figure 1.1. In the case of deterministic stationary laminar flow, all subsequent fluid elements starting from the same position follow the same path and there would be no interest in solving the flow equation for more than one initial time. However, in the case of a stochastic model of the flow, even if the macroscopic conditions are stationary, at a microscopic level the flow path is unstable and fluid elements departing from the same position at different times can follow completely different paths. In this case then, the different realizations can be taken to represent these different paths. Alternatively, referring again to Figure 1.1, one can consider different fluid elements that all start from the same X-coordinate at the same time, but at different positions on the cross section perpendicular to the X-axis of the porous medium. If we track the x- coordinate of each of these fluid element as a function of time, these will clearly develop differently for each fluid element and once more it would be reasonable to consider them each as a different realization of a 1-dimensional flow equation that models flow along the X-direction.

Whether the realizations are distinguished according to starting times or starting positions, the details of the movement of a single fluid element is clearly not of physical interest, but rather the effect of superimposing all the fluid elements. If in the example above a contaminant is introduced into the flow at a certain point, the concentration profile further downstream is determined by adding the contributions from each of the fluid elements that received some contaminant at the injection point, but then followed different paths and are found at different X positions after a given time lapse.

To model the macroscopic behavior of contaminant transport, it would therefore be very desirable to find a direct mathematical description of the temporal behavior of the statistical quantities, such as the mean and standard deviation of the position, rather than having to extract this from multiple simulated individual realizations.

It turns out that there is indeed a way to achieve this improbable goal, and that is the subject of this chapter. It is based on a major development in the theory of stochastic differential equations that was performed over the second half of the 20th century by many of the leading mathematicians of the time, and the discussion here will only aim to convey the essentials of that theory in a way that explains its application to the porous medium flow problem.

The outcome of the theory is that the problem of solving a particular SDE is replaced by that of solving a connected deterministic differential equation, of the sort that is well known in potential theory. The concept of a potential arises in physical problems in many areas, such as electrostatics, hydrodynamics and thermal conduction. A typical problem from the latter area will serve as a straightforward example. Consider a heat conductor in which there is a temperature gradient because it is located between a heat source and a heat sink, but the system is in a steady state so that the temperature everywhere remains constant with time. The basic law of heat conduction states that heat flow is proportional to the gradient of the temperature, which means that temperature T plays the role of a potential for heat flow. It is easily proven from the conduction law that inside the conductor, where there are no heat sources, T must satisfy the partial differential equation

$$\nabla^2 T = 0 . \tag{5.1}$$

This is the Laplace equation, that also applies, for example, to the electrostatic potential in a charge-free region. For any point P inside the conductor, it is intuitively plausible that the temperature at P is equal to the average of the temperature over any spherical surface within the conductor and centered on P. If , for example, the average on this sphere was higher than at P, heat would flow from it towards P, and the temperature would no longer be stationary.

The statement can also be mathematically proven to be a direct consequence of the Laplace equation and is often called the *mean value theorem.* The problem of solving the Laplace equation subject to known values of the potential on an enclosing boundary, is usually referred to as the Dirichlet problem in potential theory.

In a seminal paper, it was pointed out by Shizuo Kakutani (1945) that there is a close connection between the Dirichlet problem and random walks. Let P be any point in a source free region of the potential Φ, and S is an enclosing boundary around P. If one considers all random walks that start at point P, and notes the value of the potential at the first time that each random walk crosses S, the average of these values over all the random walks approaches a statistical expectation value $E[\Phi(S)]$ as the number of walks tends to infinity. It was proven by Kakutani that

$$E[\Phi(S)] = \Phi(P) . \tag{5.2}$$

The proof is quite straightforward, and relies on the mean value theorem as well as the Markov property of random walks (i.e., that the subsequent development of a random walk is independent of the way in which a particular point on the path was reached). The proof is discussed on a non-mathematical level by Hersh and Griego (1969).

In the special case that S is chosen to be a sphere, it is easy to see that equation (5.2) merely reduces to the mean value theorem. As the incremental displacements in the random walk are equally likely to be in any direction, all points on a spherical S are equally likely to be the exit point of the random walk. Hence a very large number of random walks sample all parts of the spherical surface equally, and the expectation value is just the ordinary average.

However, for any other shape of the surface, equation (5.2) is a generalisation and may in fact yield a practical method of solving the Dirichlet problem approximately in cases where the shape of surface on which the boundary values are known is too complicated to allow computation directly from the deterministic differential equation of the potential.

In subsequent development of Kakutani's basic idea by many authors, a comprehensive mathematical framework has been established that moves beyond the Laplace equation and random walks. It leads to a well-defined correspondence between a large class of stochastic differential equations and an associated set of deterministic 2nd order partial differential equations. No attempt will be made here to present all the theory and derivations; these are well covered in the books by Øksendal, and Kloeden and Platen. We will restrict ourselves to introducing some definitions and stating the theorems, together with a discussion of simple illustrative examples to demonstrate the applications to the flow problems that is our main concern.

5.2 Ito Diffusions

The theory that follows is restricted to SDE's of the type

$$dX_t = b(X_t)dt + \sigma(X_t)dB_t \,. \tag{5.3}$$

Note that the so-called *drift coefficient* b and *diffusion coefficient* σ are restricted to be independent of the time variable, unlike the more general case represented e.g. by equation (2.4). Moreover, b and σ must satisfy a continuity condition (the Lipschitz condition):

$$|b(x)-b(y)|+|\sigma(x)-\sigma(y)| \leq D|x-y| \tag{5.4}$$

where D is an arbitrary constant, for all x and y. An SDE that satisfies these restrictions is called an *Ito diffusion* (ID). Note, however, that X_t and B_t are not restricted to be scalars – they can be n- and m-dimensional vectors respectively, and b and σ correspondingly a vector and a matrix respectively, in n x m dimensions. So equation (5.3) really represents a set of first order SDE's rather than a single equation. Moreover, this means that an ID is not restricted to 1st order equations only, since higher order differential equations can be reduced to a set of 1st order equations by introducing derivatives as independent variables.

The term "Ito diffusion" is a mathematical convention that can become confusing in a study of actual flow and transport phenomena, where real physical diffusion is relevant. We therefore prefer to only use the acronym "ID" to identify the mathematical concept. The simplest concrete example of an ID, is the 1-dimensional Brownian motion itself, given trivially by

$$dX_t = dB_t\,. \tag{5.5}$$

The same formal equation would represent 2-dimensional Brownian motion, provided that we interpret it as an equation for the vectors

$$X = \begin{pmatrix} X_1 \\ X_2 \end{pmatrix}, \quad B = \begin{pmatrix} B_1 \\ B_2 \end{pmatrix}; \qquad b = (0,0), \quad \sigma = \begin{pmatrix} 1 & 0 \\ 0 & 1 \end{pmatrix}. \tag{5.6}$$

Another interesting example of an ID is the pair of SDEs

$$\begin{aligned} dY_1 &= -\tfrac{1}{2}Y_1 dt - Y_2 dB, \\ dY_2 &= -\tfrac{1}{2}Y_2 dt + Y_1 dB, \end{aligned} \tag{5.7}$$

which is the special case of equation (5.3) where

$$b = \begin{pmatrix} -\frac{1}{2}Y_1 \\ -\frac{1}{2}Y_2 \end{pmatrix}, \quad \sigma = \begin{pmatrix} 0 & -Y_2 \\ Y_1 & 0 \end{pmatrix}, \quad \mathbf{B} = \begin{pmatrix} B \\ B \end{pmatrix} = B\begin{pmatrix} 1 \\ 1 \end{pmatrix}. \tag{5.8}$$

Notice that although B in this equation is a vector, since its components are equal there is only one independent stochastic term. The solution to equation (5.7) satisfies

$$\begin{aligned} dY_1(t) &= -\sin(B)dB - \tfrac{1}{2}\cos(B)dt, \\ dY_2(t) &= \cos(B)dB - \tfrac{1}{2}\sin(B)dt, \end{aligned} \tag{5.9}$$

as is easily proven from the Ito formula by applying it to the vector function $g(t,x) = exp(ix) = (cos\ x,\ sin\ x)$. From equation (5.9) it is clear that if the initial values of the vector $Y = (Y_1\ ,\ Y_2)$ is chosen to fall on the unit circle, Y will always stay on the unit circle. This ID can be called *Brownian motion on the unit circle* and in this case the B can be interpreted as a stochastically varying angle coordinate.

5.3 The Generator of an ID

The next important step is to define the generator of an ID. The generator A is a special kind of differential operator, defined by the following limit:

$$A\,f(x) = \lim_{\Delta t \to 0} \frac{E^x\left[f(X_t)\right] - f(x)}{\Delta t} \,. \tag{5.10}$$

Here, f is an arbitrary function of the n-dimensional vector x, and the notation E^x means the expectation value taken over all realizations of the ID X_t that start at $t = 0$ at the particular position x. In words, A gives the expected time rate of change of a function, as its argument undergoes an Ito diffusion away from the starting point at x.

It is proven in Øksendal (1998) (theorem 7.3.3) that for a given ID, the associated A can be expressed in terms of the coefficient matrices of the ID equations as follows, provided that f is twice differentiable:

$$A\,f(x) = \sum_i b_i(x) \frac{\partial f}{\partial x_i} + \tfrac{1}{2} \sum_{i,j} (\sigma \cdot \sigma^T)_{ij}(x) \frac{\partial^2 f}{\partial x_i \partial x_j} \,. \tag{5.11}$$

Notice the appearance of the second derivative, although equation (5.10) ostensibly defines A as a first derivative. This is a manifestation of the inclusion of 2nd order differentials of the Wiener increment that was discussed in connection with equation (2.5). The proof of equation (5.11) involves using the Ito formula to express the differential of f ; when this is integrated, one term in the result takes the form of an Ito integral, but this term falls away when the expectation value that appears in equation (5.10) is taken and the result is as shown above.

In the literature, a somewhat generalized definition of the generator is sometimes encountered and often referred to as the *characteristic operator* of the ID. However, as shown in Øksendal (1998) wherever the generator exists the two operators are identical and equation (5.11) is a valid representation of both operators.

5.4 The Dynkin Formula

In integrating a deterministic differential equation for some function f in the time domain, one may usually integrate from $t = 0$ to some indefinite final time t. The solution found in this way, will directly answer a question such as about the value(s) of t at which a given value of f will be reached.

In the case of an SDE, the situation is more complex. While one could still integrate the SDE up to a chosen time, each realization of the solution will reach a predetermined value at a different time, so no unique t value corresponds to a particular value of f. To clarify this, the concept of a *stopping time* is introduced. A stopping time τ is defined as a fixed time value for which it is possible to decide on the grounds of a single realization whether τ has been reached. Any specified time value is a valid stopping time, and so is the time at which a specified value of f is reached for the first time. The latter is an example of a *first exit time*; if a subset of realizations is defined by a condition that can be applied to any realization at time t to decide its membership of the subset, the first exit time from this subset is a valid stopping time. By contrast, the expected time for f to reach a specified value is not a valid stopping time, because it is a statistical quantity that requires knowledge about all possible realizations.

Using this concept we can now formulate Dynkin's formula, which reads:

$$E^x[f(X_\tau)] = f(x) + E^x\left[\int_0^\tau A f(X_s)\,ds\right]. \tag{5.12}$$

Here, τ is a stopping time satisfying $E^x[\tau] < \infty$ while f is as before any twice differentiable function.

The derivation of Dynkin's formula is contained in the derivation of equation (5.11) as outlined above. It is also intuitively plausible as essentially the expectation value of an integral of equation (5.10). Despite this apparently straightforward origin, it has profound consequences and is the basis of most of the SDE theory that we have applied in the flow problem. An extended account of this result can be found in Dynkin (1965).

If we consider Dynkin's formula for the case that τ is a fixed time t so that the expectation value on the right hand side of equation (5.12) can be taken inside the integral, and define

$$u(t,x) = E^x[f(X_t)]. \tag{5.13}$$

We may differentiate equation (5.12) to find

$$\frac{\partial u}{\partial t} = Au, \qquad t > 0 \tag{5.14}$$

$$u(0,x) = f(x). \tag{5.15}$$

and moreover, as proved in Øksendal (1998) Chapter 8, the reverse also holds: for any appropriately differentiable function that solves equations (5.14) and (5.15), equation (5.13) also holds. This differential version of Dynkin's equation, sometimes referred to as Kolmogorov's backward equation, is in some cases simpler to apply than the original integral version.

5.5 Applications of the Dynkin Formula

We first illustrate that Dynkin's equation is a generalisation of Kakutani's result that was discussed in the introduction, section 0. Suppose the ID under discussion is simple 3-dimensional Brownian motion, represented by equations such as (5.5) and (5.6). The generator of this ID is found by putting the 3-dimensional equivalents of equation (5.6) into equation (5.11) yielding

$$Af = \nabla^2 f. \tag{5.16}$$

We choose a bounded region S in space surrounding the starting point x of the Brownian motion, and define τ to be the first exit time from this region. Then, if f is chosen as a solution of the Laplace equation inside S, and equation (5.16)(5.16) substituted into Dynkin's equation, the integral on the right vanishes and Kakutani's result follows.

A very interesting result is obtained if we choose S to be an annular region bounded by a small inner sphere of infinitesimal radius ε centered on an arbitrary point b, and an outer concentric sphere of radius $R >> \varepsilon$, chosen big enough that the starting point x is inside the annulus. The stopping time is taken as the first time the Brownian motion exits S across either boundary. Let p be the probability that it leaves S across the inner boundary first, i.e. that it reaches point b, and $q = 1\text{-}p$ the probability that it leaves S across the outer boundary.

We choose f to be spherically symmetric around point b, i.e. only a function of the radius measured from b. This is a choice allowed by the Laplace equation. The expected value of f at the stopping time is just the probability weighted sum of f values on the two boundaries. If we choose the boundary condition $f(\varepsilon) = 1$ and $f(R) = 0$, it follows that

$$E^x[f(X_\tau)] = p. \tag{5.17}$$

We consider this problem for the case of n spatial dimensions. Expressing the Laplace operator in the appropriate radial coordinates according to the number of dimensions, the solution satisfying the stated boundary conditions is easily found by direct integration to be

$$\begin{aligned} f(r) &= \tfrac{R-r}{R-\varepsilon} &&; n=1 \\ f(r) &= \tfrac{\ln(R)-\ln(r)}{\ln(R)-\ln(\varepsilon)} &&; n=2 \\ f(r) &= \frac{\frac{1}{R}-\frac{1}{r}}{\frac{1}{R}-\frac{1}{\varepsilon}} &&; n=3 \end{aligned} \tag{5.18}$$

Applying the Kakutani result equation (5.2) to equation (5.17)(5.17), it follows that $p = f(r_0)$ where $r_0 = |x\text{-}b| > \varepsilon$, is the starting radius. Consider now the effect of relaxing the restriction to the finite spatial region enclosed by S, by taking the limit as $R \to \infty$. From equation (5.18) it is seen that in the cases of 1 and 2 dimensions, $p \to 1$, but for 3 (or more) dimensions $p \to 0$ (as $\varepsilon \to 0$). This means that in 1 or 2 dimensions we can be sure, in a probabilistic sense, that starting from an arbitrary spatial point, Brownian motion will eventually reach any other arbitrarily chosen point (in the example, the point b chosen as the centre of the annulus); but in more dimensions, this probability vanishes. The argument is easily extended to say that in less than 3 dimensions, a Brownian motion starting from a given point will eventually return to the point, i.e. it is recurrent; but in 3 or more dimensions it is not recurrent.

This result is known as Polya's theorem. The power of the Dynkin formula is demonstrated by the ease by which this subtle result was obtained, compared to the original proof by Polya (1921).

It is a feature of the application of Dynkin's equation, that we do not usually have a preconceived notion of the function f for which it allows us to calculate the expectation value. Instead, we start by finding the generator from the SDE that describes a problem. Then we construct a differential equation that f should satisfy in order to simplify the integral on the right hand side of Dynkin's equation – such as equation (5.16) above. In this way Dynkin's equation dictates the form of the function for which expectation values are found, and this form is different for different ID's and even for the same ID, different forms are obtained depending on how the right hand side of the generator equation is chosen. That is demonstrated by the example in the next section.

5.6 Extracting Statistical Quantities from Dynkin's Formula

The procedure by which statistical properties of the solutions to an SDE can be extracted from Dynkin's formula, will now be demonstrated by applying it to the population growth problem.

The first step is to find the generator for the SDE, contained e.g. in equation (2.4). Identifying the drift and diffusion coefficients of equation (5.3) (5.3) as $b(x) = r\,x$ and $\sigma(x) = \sigma\,x$ equation (5.11) becomes:

$$A\,f(x) = rx\frac{\partial f}{\partial x} + \tfrac{1}{2}\sigma^2 x^2 \frac{\partial^2 f}{\partial x^2}. \tag{5.19}$$

In order to be useful in Dynkin's formula, an f is now to be found such that it makes the integral on the right hand side of the equation tractable. The simplest choice is to find f such that $A\,f = 0$; that was the choice which led to Kakutani's theorem in the previous section. Other possibilities are to make $A\,f$ = *constant* or $A\,f = \partial f/\partial t$. Each choice supplies the answer to a different question about the SDE solution and will be discussed separately below.

5.6.1 What is the probability to reach a population value K ?

To answer this, we define the stopping time as the first time the population exits from the bounded set of values defined by the interval [ε,K] or in other words, the open interval (0,K] . The value 0 is excluded because it can never be reached, as is also clear from the explicit solution in equation (2.5). Using the choice $A\,f = 0$ and equation (5.19) it is easily seen that

$$f(x) = \frac{x^{\gamma}}{\gamma}\;;\; \gamma = 1 - \frac{2r}{\sigma^2}, \tag{5.20}$$

where γ is a dimensionless constant. At the stopping time, X_t must either have the value ε or K; define p_ε and p_K respectively as the probabilities for each of these to happen. Dynkin's equation reduces to

$$E^{X_0}[f(X_\tau)] = p_\varepsilon f(\varepsilon) + p_K f(K) = f(X_0) + 0 \tag{5.21}$$

and by using $p_\varepsilon + p_K = 1$ and equation (5.10) it is found that

$$p_K = \frac{X_0^{\gamma} - \varepsilon^{\gamma}}{K^{\gamma} - \varepsilon^{\gamma}}. \tag{5.22}$$

The case of interest is when $\varepsilon \to 0$ for which two expressions are obtained:

$$p_K \xrightarrow[0]{\varepsilon} \begin{cases} \left(X_0 / K \right)^{\gamma} & ,\gamma > 0 \\ 1 & ,\gamma \le 0 \end{cases} \tag{5.23}$$

The deterministic limit is when $\sigma = 0$ i.e. when $\gamma \to -\infty$. In this case any finite value $K > X_0$ will always be reached as the population is growing exponentially, and equation (5.23) shows that the same is true for any negative γ. However, for strictly positive γ, there is a finite probability that K will never be reached and this increases as either γ or K increases.

This straightforward result would be quite difficult to obtain from numerical simulations; especially when the probability is low, a very large number of simulations would be needed to obtain reliable statistics. Moreover, the result shows that there is a marked qualitative difference in the behavior below and above the critical value $\gamma = 0$ that separates the regions where the deterministic and stochastic contributions dominate. This fact is not so clear even from the explicit stochastic solution of equation (2.4).

5.6.2 What is the expected time for the population to reach a value K?

This question is answered by choosing $A\,f =$ *constant*. The constant must have the dimensions $[T^{-1}]$ because as is clear from equation (5.10), A is essentially a time derivative. We may therefore simplify the calculation by choosing the constant to be r, since it is the only relevant constant in the problem and has the right dimension. Solving the resulting generator equation yields

$$f(x) = \frac{x^{\gamma}}{\gamma} + \frac{\gamma - 1}{\gamma} \ln x . \tag{5.24}$$

Dynkin's equation now becomes

$$E^{X_0}[f(X_\tau)] = f(X_0) + E^{X_0}\left[\int_0^\tau r\,ds\right] = f(X_0) + rE^{X_0}[\tau] = f(X_0) + rT . \tag{5.25}$$

where T is the expected exit time. Once more expressing the expectation value on the left hand side in terms of the probabilities, substituting equations (5.25) and (5.22) and simplifying one obtains

$$rT = \tfrac{\gamma-1}{\gamma}[p_K \ln K - \ln X_0 + (1 - p_K)\ln \varepsilon] . \tag{5.26}$$

The deterministic limit of this ($\gamma \rightarrow -\infty$ and $p_K \rightarrow 1$) gives a time that is identical with that solved directly from the deterministic exponential growth solution. However, the behavior of the stochastic solution is most easily interpreted from a numerical plot (see Figure 5.1).

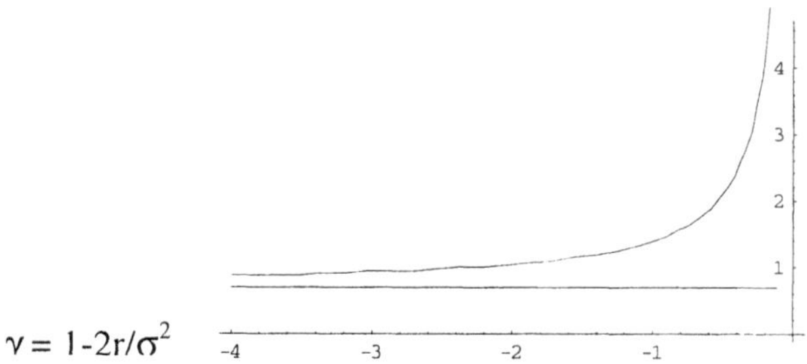

Figure 5.1 The expected time for the population to double as stochastic amplitude increases.

Figure 5.1 shows the expected time for the population to double, i.e $K=2\ X_0$. The straight line is the deterministic time, and as shown this is approached as $\gamma \to -\infty$. However, for the stochastic solution the expected time is always more than the deterministic value, and it becomes infinite for $\gamma \geq 0$.

To put it in perspective, we note that the realizations explicitly calculated in Chapter 3 represent a value of $\gamma = -2$ for Figure 3.13, and $\gamma = -11$ for Figure 3.14 respectively. In both these cases the behavior remains essentially one of unlimited growth, although an extended time is needed to reach a population doubling. However the implication is that in the region where stochastic contributions dominate, the population of a "typical" realization will never reach the value K.

This result is a stronger one than that of the previous subsection, which was that there is a finite probability that K will never be reached. Here we find that even where there is a finite probability that K will be reached, one may on average have to wait an infinite time for this to happen. These statements may appear contradictory; but before discussing it, we first address the next question since it appears to throw up an even stronger contradiction.

5.6.3 What is the Expected Population at a Time t ?

One possibility to answer this question is to extend the treatment to a 2-dimensional problem, in which time becomes the second variable, represented as a trivial "stochastic" variable regulated by the equation $dX_2 = dt$, i.e. without an actual stochastic variation. The expected population can then be found by taking the stopping "time" as the escape from a rectangular region bounded by the lines $X_1 = K$ and $X_2 = T$ in the (X_1, X_2) plane. An example of this approach is found in Øksendal(1998).

However, we show a more direct technique based on the Kolmogorov backward equation (5.14). This corresponds to the case where we put a partial time derivative on the right hand side of the generator equation, and it becomes a partial differential equation in the two variables x and t. Using the same generator as in the previous subsections and solving the equation by separation of variables introducing a separation constant C, the solution is:

$$u(x,t) = e^{\frac{Cr}{1-\gamma}t} x^{\frac{1}{2}[\gamma+\sqrt{\gamma^2-4C}]} . \tag{5.27}$$

This is a solution for all values of the separation constant; we choose C such that the exponent of x becomes an integer N, in terms of which equation (5.27) is expressed as

$$u(x,t) = x^N e^{Nrt+\frac{1}{2}N(N-1)\sigma^2 t} . \tag{5.28}$$

If we now apply equations (5.15) and (5.13) we find that

$$E^{X_0}[X_t^N] = (X_0 e^{rt})^N e^{\frac{1}{2}N(N-1)\sigma^2 t} . \tag{5.29}$$

A case of particular interest is the mean value μ of the population, i.e. $N = 1$. Using this, the case $N = 2$ yields a value for the standard deviation s as well:

$$\mu = X_0 e^{rt} ; \qquad s = X_0 e^{rt} \sqrt{e^{\sigma^2 t} - 1} . \tag{5.30}$$

Notice that the functional form of the function of x for which the expectation value is given by Dynkin's formula, is dictated by the solutions to the generator equation. Different forms were obtained in answering the various questions above. In the first cases examined, individual functions were obtained, namely equations (5.20) and (5.24). In the case of the expected population, instead, a whole family of functions was found, i.e. integer powers of x, making the extraction of statistical moments particularly simple. For other problems, more mathematical manipulation is usually required to achieve that.

A noteworthy feature of equation (5.30) is that μ is given by exactly the deterministic expression for the population, irrespective of the value of σ. While this behavior would be expected if the stochastic solution merely consisted of a random variation superimposed on the deterministic behavior, the results of the previous subsections showed that in fact the behavior of the stochastic solution is more complex. In fact, the behavior of μ seems difficult to reconcile with that of the probability and the expected time – if the average value grows exponentially with time, how can it take an infinite expected time (for some range of σ values) to reach a fixed value, such as twice the initial population?

The resolution of this paradox lies in the fact that the population value is bound from below, as the exponential growth does not allow it to reach 0, but not from above. At any fixed time, there must be some realizations which have populations far above μ. To balance these and restore the average, there must be many more realizations in the range between 0 and μ. So the majority of realizations actually have populations below average, and this accounts for the fact that the expected time to reach a fixed value is always larger than the deterministic time. When the stochastic term dominates, the probability of reaching the fixed value is less than one according to equation (5.23), meaning that a finite fraction of realizations never reach it. The fraction that does not reach it in a finite time interval must be even bigger.

For the deterministic system, the questions about the population at a given time, and about the time to reach a fixed value, are two sides of the same coin in the sense that their answers can be read off from the same point on a plot of population vs. time.

The stochastic term destroys this relation. In fact, the argument above suggests that the behavior of the population mean, while simple, is not representative of the behavior of a typical realization. Conversely, if a finite number of realizations are generated numerically, the mean value would not give a reliable estimate of the population mean, because the sample would not be likely to include enough of the low probability, high population realizations. In this case the estimated time to reach a fixed value, would give a more realistic description of the behavior of typical realizations.

This discussion highlights the fact that SDE solutions can be inherently different from the underlying deterministic system, and do not just represent a random variation about a deterministic mean value. This is particularly visible in the population growth example, because of the highly non-linear properties of the exponential growth curve.

The advection-dispersion approach to contaminant transport discussed in Chapter 1 was in fact just such a simplification based on superimposing random variation on a deterministic transport equation. It may be that the solutions to the flow equations have a more moderate behavior than for exponential population growth, giving more practical justification of the a priori splitting of the velocity into a mean drift velocity and fluctuations. However, in principle the objection that stochastic variation needs to be included into the differential equation itself for a valid description remains valid.

5.7 The Probability Distribution of Population Growth Realization

The calculation of the previous section contains far more information than just the mean and standard deviation of the population. In fact, equation (5.29) specifies the value of all statistical moments of the probability distribution, and as is well known this is enough to fully determine the distribution itself. If an explicit expression for the distribution can be derived, this will facilitate the calculation of any other statistical quantities of interest.

The problem can be mathematically formulated by rewriting equation (5.29) as:

$$\int_0^\infty dx\, x^N P(x,t) = (X_0 e^{rt})^N e^{\frac{1}{2}N(N-1)\sigma^2 t} \tag{5.31}$$

where $P(x,t)$ is the desired probability density of finding a population x at time t. The appearance of the factor x^N in the integrand is a consequence of the structure of the generator for the population growth SDE. Other problems will yield a similar integral on the left of the equation, but with different factors in the integrand. A common feature, however, is that while the Dynkin equation refers only to a single function $f(x)$ for which the expectation value is calculated, the solution of the time-dependent equation (5.14) in fact yields a family of solutions, indexed by the value of an integration constant such as the separation constant that manifests in equation (5.33) as the integer value N. Such a more general form of the equation could be written as

$$\int f_n(x) P(x,t) = u_n(x_0,t)\,. \tag{5.32}$$

While there is no unique solution for $P(x,t)$ from this integral equation for a single known function $f(x)$ and $u(x,t)$, such a unique solution does exist if the equation is satisfied by each member of a linearly independent set $f_n(x)$. In fact, the solution is facilitated if the $f_n(x)$ form a complete orthonormal set of functions, as is often the case when the set is found as solutions of a differential equation. If so, one may use the completeness relation which is typically of the form

$$\sum_n f_n(x) f_n(x') = \delta(x-x')\,, \tag{5.33}$$

where $\delta(x)$ is the Dirac delta-function, to solve equation (5.32):

$$P(x,t)=\sum_{n} f_n(x)u_n(x,t)\,. \tag{5.34}$$

Equation (5.31) cannot be solved in this way directly, because the set of functions x^N are linearly independent but not orthonormal. However, each x^N can be expressed as a linear combination of any of the well-known sets of orthogonal polynomials such as Hermite or Laguerre polynomials, and in this way a similar solution for *P(x,t)* in the case of the population growth problem can be constructed.

While the problem of calculating the probability distribution is formally solved this way, it turns out that the resulting infinite sum expression suffers from convergence problems in practical numerical calculations. This is closely connected with the fact that the infinitely sharp peak on the right hand side of equation (5.33) cannot be successfully approximated by any finite sum of orthogonal polynomials. Consequently only imperfect representations of the probability density valid for limited ranges of the arguments are obtained. It turns out that in cases of interest in the contaminant flow problem, more elegant but ad hoc solutions for the probability density can be constructed. We will therefore not pursue the details of the solution for the population growth problem any further, beyond pointing out that even for this relatively simple problem, solving for the probability density is in principle possible but in practice quite difficult.

Chapter 6

Stochastic Modeling of the Velocity

6.1 Introduction

The purpose of developing the theory of stochastic processes and stochastic differential equations (SDE's) in Chapters 2-4, was to apply it to the problem of random variations of the flow velocity on a microscopic scale, for a fluid flowing through a porous medium. The macroscopic flow is adequately represented by Darcy's equation; an obvious idea is to represent the flow velocity by augmenting this with random fluctuations such as represented by a white noise term:

$$\underline{V}(\underline{x},t) = -\frac{K(\underline{x})}{\varphi(\underline{x})}\nabla\phi(\underline{x},t) + \underline{W}(\underline{x},t)\,. \tag{6.1}$$

Here, as before, K is the hydraulic conductivity, φ the porosity, and ϕ the hydraulic pressure, also called the piezometric head. This equation shows the interrelatedness of the mean velocity and the noise component, and the Darcian term on the right hand side can be replaced by a known function for a particular physical situation.

Equation (6.1) is not a differential equation, but it becomes one if the velocity is expressed as the derivative of the position vector of a fluid element. Equation (6.1) then assumes a form very similar to that of equation (2.4). One can envisage that in a similar way as in chapter 2, the requirement that the displacement of the fluid element must be continuous even if subject to random increments, will eventually lead to an SDE similar to equation (2.4), in which the random variations are represented by a Wiener process.

However, to actually do this requires some extensions of the theory as developed so far. Firstly, equation (6.1) is a vector equation in which the velocity, position and the noise term are all spatial vectors in 1,2 or 3 dimensions. This is not a major concern, as by writing it in components the single equation becomes a set of equations all of the same form and although the examples discussed in Chapters 2-4 all contained only a single equation, the theoretical expressions do apply to sets of linear equations as was briefly discussed after equation (6.4).

More importantly, the random term is shown in equation (6.1) as depending on both space and time coordinates and consideration has to be given to the nature of the random dependence on more than one variable. This is further explored in the next section. Note that it is, indeed, physically plausible to expect random variation associated with both position and time. In an experiment such as that by Rashidi et al. (1996), which was discussed in section 1.2.3, if a snapshot is taken of the velocities at different positions, these vary randomly about the average drift velocity. Conversely, if one focuses on one position, and registers the velocity as a function of time, random variation is generally observed indicating that on the microscopic scale the flow is not laminar. From another perspective, if the velocity is interpreted as that of a particular fluid element, its instantaneous velocity will undergo random fluctuations over time as it progresses along its trajectory, because of the pore structure that it encounters along the way.

One may also reflect on the physical interpretation of the random variation introduced here, in the light of the fact previously stressed that an SDE arises when the driving coefficients in a differential equation varies stochastically. In Darcy's equation, expressed as a differential equation for the displacement, the driving coefficient is derived from the medium properties K and φ, and an external system property ϕ. Is it reasonable to use random variations in any of these to produce random microscopic fluid displacements? An argument is most easily made out for the hydraulic conductivity K. As formulated above, K is shown as a scalar quantity that in principle indicates a displacement of the fluid element in the direction of the pressure gradient. However, more generally K is a tensor and its off-diagonal elements represent displacements in other directions than that of the pressure gradient that drives them. This is exactly the effect that the pore structure of the medium has on the microscopic velocity: a fluid element that strikes a grain wall, is deflected away from the direction of the external pressure gradient, and in turn will also affect the directions in which neighboring fluid elements travel. Thus it is entirely reasonable to imagine replacing the physical obstacles that the porous medium offers to the flow, by fluctuations in the conductivity tensor. Conceptually then, one may contemplate using a full tensor conductivity, and put random variations into each of its components; but as the net result will simply be to produce random displacements away from that produced by the scalar macroscopic tensor, the same effect is reached by directly introducing the random displacements themselves. That is essentially what is done when using equation (6.1) in the stated form.

6.2 Spectral Expansion of Wiener Processes in Time and in Space

If we replace the white noise term in equation (6.1) by a Wiener process that is random with respect to both space and time, this will have severe consequences for numerical modeling. As illustrated in Chapter 3, a discrete grid needs to be introduced for each independent variable; if we choose N grid points for each, this will give N^2 grid points for a one dimensional problem, or N^4 grid points for a problem in 3 dimensions. At each grid point, a random Gaussian value must be calculated.

However, it may be questioned if this amount of randomness is meaningful. The goal to make a fluid element follow a random path through the porous medium is equally well achieved by adding a Wiener process that only varies stochastically with either time or space alone. Suppose that we add a time dependent Wiener process. As the fluid element proceeds along its path, a random displacement will be added to its instantaneous motion at every point along the path even if there is no explicit spatial randomness. Also, a subsequent fluid element that starts from an identical initial position and velocity, will follow a different random path. This is a plausible representation of the non-laminar aspect of the microscopic flow. The only aspect that would not be plausible with a purely time-dependent Wiener term, is that if we could take a snapshot of the random displacements of all fluid elements at one instant, these would all be the same. In other words, the displacements would be perfectly correlated in space over the entire extent of the medium. This is not physically plausible; however, a spatial correlation over small distances would not only be acceptable but would in fact add to the realism of the stochastic model. For example, all the fluid elements within the volume of a single pore can clearly not move independently and their motions should be correlated.

The conclusion is that if we add a stochastic term that is a Wiener process with respect to the time variable, but is spatially correlated over a finite range (the correlation length, b), this will give a model that is both physically reasonable and numerically tractable. The remaining question is how to formulate such a stochastic term mathematically. To answer that, it is first necessary to explore some properties of the particular kind of stochastic process defined below. The treatment presented here, is based largely on that discussed by Ghanem and Spanos (1991).

Consider an arbitrary set of functions $\{f_n(x)\}$, that is orthonormal over some domain D; i.e:

$$\int_D f_n(x) f_m(x) dx = \delta_{nm} . \tag{6.2}$$

We also select a set of random variables, $\xi_n(\omega)$ where as in Chapter 2, ω is a label that identifies a particular value of the n-th random variable. The ξ_n are restricted to have a zero mean over all values of ω, for each fixed index n. Furthermore, they are statistically independent so that they satisfy

$$E[\xi_n(\omega)\xi_m(\omega)] = \delta_{nm} . \tag{6.3}$$

Then the sum

$$\alpha(x,\omega) = \sum_{n=0}^{\infty} \sqrt{\lambda_n} f_n(x) \xi_n(\omega) \tag{6.4}$$

constitutes a stochastic process; e.g., for each value of ω, the values taken by $\alpha(x,\omega)$ is a realization of the process α. The λ_n are numerical coefficients, whose values will be specified below. By taking the mean of equation (6.4), it follows that α is a process with zero mean value. A process defined in this way has an implied correlation between values at different values of its independent argument, as is shown by calculating the covariance. Because it is a zero-mean process, the covariance function $C(x_1,x_2)$ reduces to

$$\begin{aligned} C(x_1, x_{2)}) &= E[\alpha(x_1,\omega)\,\alpha(x_2,\omega)], \\ &= \sum_{n=0}^{\infty}\sum_{m=0}^{\infty} \sqrt{\lambda_n \lambda_m} f_n(x_1) f_m(x_2) E[\xi_n(\omega)\xi_m(\omega)]. \end{aligned} \tag{6.5}$$

Applying equation (6.3) removes one summation in the second line of equation (6.5):

$$C(x_1, x_2) = \sum_{m=0}^{\infty} \lambda_m f_m(x_1) f_m(x_2) \tag{6.6}$$

which will not, in general, be zero, hence proving the existence of correlation. Multiplying this equation through by $f_k(x)$ and integrating over D leads, by use of the orthonormality condition equation , to the equation

$$\int_D C(x_1, x_2) f_k(x_2) dx_2 = \lambda_k f_k(x_1) . \tag{6.7}$$

Equation (6.7) has the form of an eigenvalue equation, where λ_k is now recognized as the eigenvalue and f_k as the eigenfunction of the integral kernel function $C(x_1,x_2)$.

To make the derivation consistent, it is necessary that equation (6.7) must have solutions forming an orthonormal set. It is well known from the theory of integral eigenvalue equations, that this is guaranteed provided that the kernel $C(x_1,x_2)$ is bounded, symmetric and positive definite. These properties are indeed valid, because $C(x_1,x_2)$ was defined in equation (6.5) as an autocovariance function. As a result, it has a complete orthonormal set of eigenfunctions and its eigenvalues are all positive real numbers.

Having derived the covariance function for a stochastic process defined by an expansion of the form of equation (6.4), we can now turn the argument around. Suppose that we are given a a functional expression for a covariance function $C(x_1,x_2)$. Then we can construct a stochastic process that is guaranteed to have the specified covariance, by performing the expansion in equation (6.4). In order to do this, the integral eigenvalue equation (6.7) must be soluble (and be solved) for the given $C(x_1,x_2)$.

Equation (6.4) plays a central role in this approach, and is known as a Karhunen-Loeve expansion or spectral expansion of the stochastic process $\alpha(x,\omega)$. The salient features of the Karhunen-Loeve expansion are that:

- the random behavior, and the functional dependence on the independent variable, are separated into factors in each term,
- this functional dependence can be considered known, being carried by a precisely specified set of orthnormal functions,
- one possibility is that a known set of orthogonal functions are chosen, in which case there is an implied covariance given by equation (6.6), and

- alternatively, the covariance function is predetermined, and the orthogonal functions calculated by solving an integral eigenvalue equation using the chosen covariance as the kernel of the equation.

The same idea is used in the flow problem to separate a deterministic spatial dependence and a stochastic time dependence by writing:

$$B(x,t,\omega)=\sum_{n=0}^{\infty}\sqrt{\lambda_n}\,f_n(x)b_n(t,\omega)\,, \tag{6.8}$$

where now $b_n(t,\omega)$ is a set of independent Wiener processes in time, of the type discussed extensively in Chapter 2. At a fixed time t, the Wiener process $b_n(t,\omega)$ reduces to a random variable with zero mean, such as ξ above. The independence of the b_n allows us to replace equation (6.3) by

$$E[b_n(t_1,\omega)b_m(t_2,\omega)]=\min(t_1,t_2)\,\delta_{nm}\,, \tag{6.9}$$

where use was made of one of the defining properties of a standard Wiener process. Using this relation in calculating the combined space and time covariance function $\mathcal{C}(x_1,t_1,x_2,t_2)$ in an analogous way to that above, the time dependent part factors out of the sum and we get

$$\mathcal{C}(x_1,t_1,x_2,t_2)=\min(t_1,t_2)\,C(x_1,x_2)\,, \tag{6.10}$$

where $C(x_1,x_2)$ is given by equation (6.6) as before and now represents the spatial covariance part. This factorization of the combined covariance only relied on the fact that the set of b_n all shared the same Wiener process time correlation, so equation (6.10) is easily generalized to the case where b_n are taken to be generalized Wiener processes defined by equation (2.45).

6.3 Solving the Covariance Eigenvalue Equation

The Karhunen-Loeve expansion enables us to construct a Wiener process with a predetermined spatial correlation, but it still remains to make the choice and solve for the corresponding orthonormal basis functions. The functional form of the covariance would be expected to depend on details of the flow within a pore and may depend on the properties and structure of the porous medium. However, for a general discussion it suffices to choose a simple form that encapsulates only a single property, that of a finite correlation length, and that will allow easy solution of the resulting eigenvalue equation. We assume only a single spatial dimension and choose a covariance kernel that decreases exponentially with separation:

$$C(x_1, x_2) = \sigma^2 e^{-\frac{|x_1 - x_2|}{b}} . \qquad (6.11)$$

Notice that the correlation length b appears explicitly in this expression, serving as a scale constant that regulates the distance over which correlation between stochastic processes at neighboring positions is extinguished. σ^2 is the variance of the covariance kernel which can be thought of as an amplitude factor. Equation (6.11) is to be substituted into equation (6.7) and the resulting integral equation solved for the set of λ_n and $f_n(x)$.

We take D to be a finite interval of interest, e.g. the left and right boundaries of the porous medium for which flow is modeled; D = [$-a,a$]. The equation is of the type classified as a homogeneous Fredholm equation of the second kind (See Morse and Feshbach 1953). It is most easily solved by conversion to a differential equation. The integration interval is split at the value $x_2 = x_1$, in order to avoid the modulus sign in the exponent:

$$\lambda f(x_1) = \sigma^2 \int_{-a}^{x_1} e^{-\frac{x_1 - x_2}{b}} f(x_2) dx_2 + \sigma^2 \int_{x_1}^{a} e^{+\frac{x_1 - x_2}{b}} f(x_2) dx_2 . \qquad (6.12)$$

Dividing this by σ^2 and differentiating with respect to x_1 gives

$$\bar{\lambda}\ f'(x_1) = f(x_1) + \left(\frac{-1}{b}\right) \int_{-a}^{x_1} e^{-\frac{x_1-x_2}{b}} f(x_2)\ dx_2$$
$$- f(x_1) + \left(\frac{1}{b}\right) \int_{x_1}^{a} e^{\frac{x_1-x_2}{b}} f(x_2)\ dx_2, \tag{6.13}$$

where $\bar{\lambda} = \lambda / \sigma^2$.

The first and third terms on the right hand side come from differentiation of the integral limits, and cancel. Differentiating the equation once more and resubstituting equation (6.12) leads to

$$\bar{\lambda} f''(x_1) = -2(\tfrac{1}{b}) f(x_1) + (\tfrac{1}{b})^2 \bar{\lambda} f(x_1). \tag{6.14}$$

This is a simple harmonic differential equation, and has the general solution

$$f(x) = A\cos(kx) + B\sin(kx) \tag{6.15}$$

where A and B are constant coefficients determined by the boundary conditions, and k is a wave number given by

$$k^2 = \frac{2b - \bar{\lambda}}{\bar{\lambda} b^2}. \tag{6.16}$$

The boundary conditions are derived from the integral equation for f, by evaluating equations (6.12) and (6.13) at the points $x_1 = \pm a$. This yields the pair of equations

$$\begin{aligned} f(a) + b f'(a) &= 0, \\ f(-a) - b f'(a) &= 0, \end{aligned} \tag{6.17}$$

and by substituting equation (6.15) into these, a pair of homogeneous equations for A and B are obtained. The condition for the existence of non-trivial solutions to the homogeneous system is that the determinant of coefficients must be zero, and this is met only by the discrete set of k-values that satisfy either of the following pair of transcendental equations:

$$\begin{aligned} 1 - bk\tan(ka) &= 0, \\ bk + \tan(ka) &= 0. \end{aligned} \tag{6.18}$$

Given numerical values of the correlation length b and the value a that specifies the extent of the spatial region of interest, each of the equations in (6.18) can be solved numerically for k. The two sets of discrete solutions are interleaved, and may be collected together by assigning a single numbering index n of which even values refer to solutions of the first, and odd values to the second equation. Through equation (6.16) the discretization of k implies that of the $\bar{\lambda}$ values, so that we end up with the discrete eigenvalues of integral equation (6.7) (for the exponential kernel) given by

$$\bar{\lambda}_n = \frac{2b}{1+b^2 k_n^{\ 2}} . \tag{6.19}$$

Finally, substituting the k_n into the homogeneous equations and solving for A and B, the corresponding eigenfunctions of the integral equation can be written as

$$f_n(x) = \begin{cases} \dfrac{\cos(k_n x)}{\sqrt{a + \frac{\sin(2k_n a)}{2k_n}}} & ;\text{n even} \\ \dfrac{\sin(k_n x)}{\sqrt{a - \frac{\sin(2k_n a)}{2k_n}}} & ;\text{n odd} \end{cases} \tag{6.20}$$

The square root factors in equation (6.20) ensure that the eigenfunctions are normalized over the x-interval $[-a,a]$. At this point the determination of the orthonormal basis functions is complete, and by substituting equations (6.19) and (6.20) into equation (6.4) the Karhunen-Loeve spectral expansion of a stochastic process with an exponential spatial covariance is fully determined.

6.4 Extension to Multiple Dimensions

Suppose that we apply the spectral expansion of equation (6.8) to a 1-dimensional flow model. For simplicity we assume that materials properties are constant, and there is a constant pressure gradient directed along the X-axis (e.g., water flowing through a homogeneous porous medium in a straight pipe slanting downwards). Then the SDE for the motion of a fluid element, subject to stochastic motion superimposed on Darcy's equation, can be written as:

$$dx = -\frac{K}{\varphi}\left(\frac{d\phi}{dx}\right)dt + \sigma^2 \sum_{n=0}^{\infty} \sqrt{\lambda_n} f_n(x) db_n(t,\omega) . \tag{6.21}$$

If we have 2-dimensional motion, it is reasonable to assume that the stochastic perturbations for the two spatial directions are independent. It is tempting to conclude that we simply need to add another equation of similar form for the y-coordinate. However, this is too simplistic. It would imply, for example, that since y does not occur in the x equation above, the correlation between x displacements decays only along the x-direction and not along y – something that is clearly implausible if the correlation is supposed to reflect physical interaction between fluid elements inside the volume of a pore.

6.5 Scalar Stochastic Processes in Multiple Dimensions

As mentioned in the introduction, there are two aspects of multi-dimensionality: the fact that the displacement is a vector, and the fact that it is a function of a vector, the position. We attend to the second factor first, by considering a scalar stochastic process defined in two dimensions. In fact, the realted discussion was formulated in general enough terms that it remains applicable if we interpret the variable x as a vector, and the domain D correspondingly as multidimensional. This means, for example, that equation (6.7) is reinterpreted for two dimensions as

$$\int C(x_1, y_1, x_2, y_2) f_k(x_2, y_2) dx_2 dy_2 = \lambda_k f_k(x_1, y_1) . \tag{6.22}$$

The straightforward generalization of the exponential covariance function used in section 5.3 with unit variances for simplicity, would be the isotropic expression

$$C(\vec{x}_1,\vec{x}_2)=e^{-\frac{1}{b}|\vec{x}_1-\vec{x}_2|}=e^{-\frac{1}{b}\sqrt{(x_1-x_2)^2+(y_1-y_2)^2}}. \tag{6.23}$$

When this is substituted into equation (6.22), the procedure that was applied in section 6.3 leads to the equation

$$\lambda\nabla^2 f(\vec{x})=-\tfrac{1}{b}\int\frac{e^{-\frac{1}{b}|\vec{x}-\vec{x}_2|}}{|\vec{x}-\vec{x}_2|}f(\vec{x}_2)dx_2dy_2+\left(\tfrac{1}{b}\right)^2\lambda f(\vec{x}). \tag{6.24}$$

Unfortunately this equation is, unlike equation (6.14), not a simple differential equation and is not very helpful to solve equation (6.22). However, progress can be made if we retrace our steps and write the 2D version of equation (6.4) in the form

$$\alpha(\vec{x},\omega)=\sum_{n,m=0}^{\infty}\sqrt{\lambda_{nm}}\,f_n(x)g_m(y)\xi_{nm}(\omega). \tag{6.25}$$

This expression results from the observation that if $f_n(x)$ and $g_m(x)$ are two orthonormal sets that form a complete basis for the 1-dimensional subspaces associated with the X-domain and Y-domain separately, their outer product forms a complete basis for the 2-dimensional space. As before the $\xi_{nm}(\omega)$ are required to be independent, and hence we find

$$C(\vec{x}_1,\vec{x}_2)=\sum_{n,m}\lambda_{nm}f_n(x_1)f_n(x_2)g_m(x_1)g_m(x_2) \tag{6.26}$$

which, by using the 1-dimensional orthonormality twice, leads to

$$\int C(\vec{x}_1,\vec{x}_2)f_k(x_2)g_l(y_2)dx_2dy_2=\lambda_{kl}f_k(x_1)g_l(y_1). \tag{6.27}$$

This equation is no easier to solve than equation (6.22), except if we make the assumption that C is separable into factors that depend on the space directions separately:

$$C(\vec{x}_1,\vec{x}_2)=C_x(x_1,x_2)C_y(y_1,y_2). \tag{6.28}$$

If this holds, the double integral in equation (6.27) factorizes and it is easy to see that it will be solved by choosing $f_k(x)$ and $g_l(y)$ to be the eigenfunctions of the 1-dimensional integral kernels C_x and C_y respectively, with corresponding eigenvalues λ_{xk} and λ_{yl} found from equations such as

(6.19). Moreover, the eigenvalues of the 2-dimensional equation are then given by a simple product

$$\lambda_{kl} = \lambda_{xk}\lambda_{yl}\,, \tag{6.29}$$

and as a result, the double summation in equation (6.26) also factorizes and reduces to an expression consistent with equation (6.28) if the eigenfunction expansions of the 1-dimensional kernels according to equation (6.6) is applied.

So by imposing equation (6.28), the 2-dimensional eigenvalue problem is reduced to solving two 1-dimensional problems and a 2-dimensional spectral expansion of the form of equation (6.25) directly follows. We now may consider how realistic this simplifying assumption is, by applying it to the exponential covariance function.

Since equation (6.23) does not factorize, the next best choice for a factorized exponentially decreasing covariance kernel would be

$$C_f(\vec{x}_1, \vec{x}_2) = e^{-\frac{1}{b}|x_1 - x_2|}e^{-\frac{1}{b}|y_1 - y_2|}\,. \tag{6.30}$$

The difference between these two is highlighted by making contour plots as functions of x_1, for a fixed reference point x_2. All points on a contour share the same degree of correlation with the reference point in the middle of the picture.

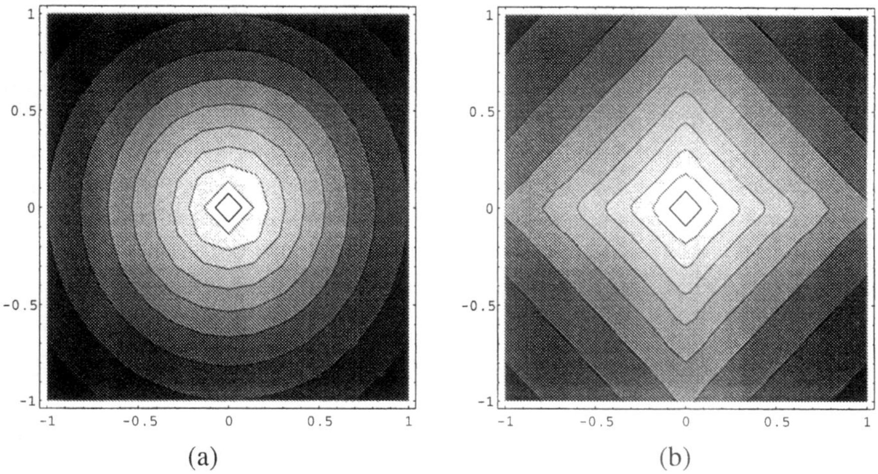

Figure 6.1 Contour maps of (a) 2D exponentially decreasing covariance function C (b) Factorised approximation C_f .

The plots in Figure 6.1 shows that the price paid for the mathematical simplification brought about by requiring C to be factorized, is a loss of symmetry. Instead of the fully isotropic correlation, it becomes only square symmetric, which is hard to justify on physical grounds. On the other hand, the choice of an exponentially decreasing covariance was itself made on the grounds of plausibility and tractability rather than a fundamental justification. Even if anisotropic, the factorized form does retain the qualitative feature of an exponential decrease at a similar rate in all directions. It therefore seems a reasonable compromise to make in a first attempt to model 2-dimensional stochastic flow.

Also note that it is possible to avoid making this approximation for some choices of the functional form of C – for example, a Gaussian choice would be naturally factorizable. In this case, the 1-dimensional problem is more difficult, however, and will not be further pursued here.

The results of this section can be extended to 3-dimensional space in a straightforward manner.

6.6 Vector Stochastic Processes in Multiple Dimensions

A vector stochastic process is obviously constructed by applying an expansion like equation (6.25) to each vector component separately. It is plausible to assume that the correlation length is the same for each vector component, and consequently to use the same sets of 1-dimensional eigenfunctions as the basis for expansion in each component, while the stochastic variations are independent for different components. Writing the vector process as $\alpha(x,\omega) = (\alpha_1,\alpha_2)$ in 2 dimensions, we have

$$\alpha_i(\vec{x},\omega) = \sum_{n,m=0}^{\infty} \sqrt{\lambda_{nm}}\, f_n(x) g_m(y) \xi_{inm}(\omega), \tag{6.31}$$

and the independence condition is

$$E[\xi_{inm}(\omega)\xi_{jkl}(\omega)] = \delta_{ij}\delta_{nk}\delta_{ml}\,. \tag{6.32}$$

The covariance is not a scalar function any more, but becomes a matrix defined by

$$C_{ij}(\vec{x}_1,\vec{x}_{2)}) = E[\alpha_i(\vec{x}_1,\omega)\alpha_j(\vec{x}_2,\omega)]\,. \tag{6.33}$$

By combining equations (6.31), (6.32) and (6.33) it is easily seen that C_{ij} is a diagonal matrix, and in fact all the diagonal elements are equal and each has the value given by equation (6.26). This means that the addition of independent stochastic variations do not introduce any cross-correlation between vector components. The deterministic differential equations for the components of a vector can each be augmented by a similar stochastic term, and only the stochastic variable in each needs to be generated independently. Therefore the case of a vector stochastic process is not mathematically significantly more complex than the scalar case, and we may directly apply equation (6.31) with the eigenfunctions and eigenvalues as found in the previous sections.

6.7 Simulation of Stochastic Flow in 1 and 2 Dimensions

6.7.1 1-D case

At this point in our discussion on the stochastic flow, it would be interesting to obtain an intuitive feel about how the covariance parameter σ^2 and the correlation length b affect the flow of a fluid particle in one dimension. The obvious way of accomplishing this would be to simulate the behavior on computer.

For the 1-dimensional case, we confine our attention to a domain D=[-1,1]. Then for the correlation function given by equation (6.11), the discrete set of k values is given by

$$\begin{aligned} 1-bk\ \tan(k)&=0, \\ bk+\tan(ka)&=0. \end{aligned} \tag{6.34}$$

A program can be written to obtain the k values for a given b, then from equation (6.19),we obtain the eigenvalue for a given σ^2:

$$\lambda_n = \frac{2\sigma^2 b}{1+b^2k_n^2}. \tag{6.35}$$

The corresponding eigenfunction can be obtained by substituting a=1 into equation (6.20):

$$f_n(x) = \begin{cases} \dfrac{\cos(k_n\ x)}{\sqrt{1+\dfrac{\sin(2k_n)}{2k_n}}} & ;\ n\ is\ even \\ \dfrac{\sin(k_n\ x)}{\sqrt{1-\dfrac{\sin(2k_n)}{2k_n}}} & ;\ n\ is\ odd \end{cases} \tag{6.36}$$

and then we use equation (6.21) to simulate the stochastic flow.

6.7.2 2-D Case

For the simple case of stochastic flow in a rectangular 2-dimensional region where the porous medium properties are constant and there is a constant pressure gradient directed along the X-axis, the 1-dimensional model of equation (6.21) is replaced by the set:

$$\begin{cases} dx = \dfrac{-K}{\varphi}\left(\dfrac{d\phi}{dx}\right)dt + \sigma^2 \displaystyle\sum_{n,m=0}^{\infty} \sqrt{\lambda_{nm}}\; f_n(x)\; g_m(y)\; db_{1nm}(f,\omega), \\ dy = \sigma^2 \displaystyle\sum_{n=0}^{\infty} \sqrt{\lambda_{nm}}\; f_n(x)\; g_m(y)\; db_{2nm}(f,\omega). \end{cases} \tag{6.37}$$

where the numeric index on the Wiener process simply identifies the x or y component to which it belongs. Equation (6.37) can be solved numerically by using, for example, an Euler scheme or a Milstein scheme which are strong Taylor approximations (Kloeden and Platen, 1992).

Chapter 7

Applying Potential Theory Modeling to Solute Dispersion

7.1 Introduction

The fundamental goal in modeling solute dispersion, is to calculate the solute concentration as a function of position and time, given an initial concentration. Chapters 2, 4 and 5 laid the groundwork by describing the mathematics of models based on stochastic differential equations. It is now time to show how these ideas can be applied concretely to calculate the evolution of the solute concentration carried by porous flow. In this chapter, we will do this for two specific cases: a 1-dimensional flow in which the carrier fluid flow velocity is constant, and a generalization of this in which the flow velocity changes linearly with position. Our emphasis will be on obtaining analytical results which can lead to insights into the relationships between flow modeling at the microscopic level, where we describe the displacement of a fluid element as a stochastic process, and the effects that this has on the macroscopic concentration. For this, the potential theory approach is the obvious choice, rather than simulation of individual realizations that will be investigated in a later chapter.

We have already formulated the stochastic model for the 1-dimensional carrier fluid flow. It is given by equation (7.1) in a slightly modified notation:

$$dx = u(x)\,dt + \gamma^2 \sum_{n=0}^{\infty} \sqrt{\lambda_n} f_n(x) db_n(t,\omega). \tag{7.1}$$

Two changes have been made to the notation. First, the combination of materials constants that appears in Darcy's law has been replaced by the function $u(x)$. This is meant not only as an abbreviation, but also to signify that the work in this chapter is applicable to any situation where a fluid velocity field can be calculated from appropriate fluid dynamics equations, not necessarily just Darcy's law. The significance of equation (7.1) is that a term has been added that models stochastic variations in this velocity field, whatever the origin of the field.

The second mainly cosmetic change is that the symbol σ that was used in equation (7.1) to indicate the variance of the covariance kernel, has been replaced by the symbol γ. This is to avoid confusion with the variance of the solute concentration, which plays an important role in the rest of this chapter. Also, it emphasizes that regardless of the way in which the factor now called γ was introduced, the role that it plays in equation (7.1) is essentially that of an amplitude characterizing the strength of the stochastic perturbation. All results of the stochastic model should reduce back to the corresponding deterministic ones in the limit $\gamma \to 0$ which we will call the deterministic limit.

As we are here dealing with the application of an ID to an actual physical problem, it is also appropriate to remind the reader that the use in equation (7.1) of Wiener processes $db_n(t,\omega)$ defined as in equations (2.5) , implies that the position and time variables have been appropriately scaled to reflect the physical processes that cause the random displacements. This was discussed in detail just after equation (2.5). In other words, x and t in equation (7.1) should really be interpreted as x/ℓ_x and t/ℓ_t where ℓ_x and ℓ_t are scale constants to be determined below. For simplicity of notation, we will suppress the scale constants until we are in a position to determine their values later on.

Equation (7.1) is already in the standard form of equation (5.3) which defines an ID. But it only specifies the fluid flow problem; to describe solute transport we have to add additional equations and the first hurdle to overcome is to formulate these equations in ID form as well.

To discuss the problem, we revert to the deterministic formulation in 3 dimensions, as the form of the equations is most familiar in that case. Equation (7.1) in that case merely becomes

$$\frac{d\underline{x}}{dt} = \underline{v} = \underline{u}(\underline{x}) \tag{7.2}$$

where $\underline{v}$ is the fluid velocity vector. The connection with solute transport is established by defining the solute flux vector $\underline{J}$ by

$$\underline{J} = \underline{v}c - D_m \cdot \nabla c \tag{7.3}$$

where c is the solute concentration and D_m the diffusion coefficient, and the evolution of the concentration is determined by solute conservation as expressed by the equation of continuity:

$$\frac{\partial c}{\partial t} = -\nabla \cdot \underline{J} \tag{7.4}$$

If we eliminate $\underline{J}$ from equations (7.3) and (7.4) we have

$$\frac{\partial c}{\partial t} = -c\nabla \cdot \underline{v} - \underline{v} \cdot \nabla c + \nabla D_m \cdot \nabla c + D_m \nabla^2 c\,. \tag{7.5}$$

This is the second equation needed in the deterministic description and for the stochastic description one would like to write this in a form that can be changed into an ID. There are a number of obstacles to this. For a start, we have to decide whether $\underline{v}$ on the right hand side of equation (7.5) is also a stochastic variable, as would be expected if it is the derivative of the position. But as the time variation of the position of a fluid element is determined by a Wiener process, it does not have a derivative!

The root of the problem is the question whether $\underline{v}$ represents the velocity of a fluid element which is at position $\underline{x}$ at time t (i.e., a particle interpretation), or is it the fluid velocity field, which e.g. in the case of stationary flow is independent of t? In changing from the deterministic equation (7.2) to the SDE (7.1) we have definitely taken the first view, but in the continuity equation we will argue below that it is the second view that is applicable.

In the case of deterministic stationary flow the distinction is unimportant. To show that, temporarily assign the superscripts "p" and "f" to distinguish between the particle and fluid interpretations. Starting with a particle interpretation of equation (7.2) we have

$$\frac{d\underline{x}^p}{dt} = u(\underline{x}^p) \quad \Rightarrow \quad \underline{x}^p = \underline{x}^p(t)\,. \tag{7.6}$$

i.e. a trajectory can be calculated for the fluid "particle". Then $\underline{x}^p$ can be differentiated to find $\underline{v}^p(t)$ and by eliminating the common parameter t we can find $\underline{v}^p(x^p)$. That gives the velocity of the particle when it arrives at the position x^p irrespective of when it gets there and in stationary flow this is the same for all particles. Then it makes sense to associate $\underline{v}^p(x^p)$ with the flow rather than the particle, i.e. $\underline{v}^p(x^p) \equiv \underline{v}^f(x)$ which is independent of time because x is now just a coordinate, independent of time. We may write

$$\frac{dv^f}{dt} = \frac{\partial v^f}{\partial t} = 0, \tag{7.7}$$

and together with the initial condition this is solved by $v^f = u(x)$, i.e we have shown that equation (7.2) applies to the flow velocity interpretation as well and there is no ambiguity about the meaning of v in equation (7.5) for the deterministic case.

However, this reasoning breaks down in the stochastic case, both because every fluid particle follows a different trajectory and because $\underline{x}^p(t)$ cannot be differentiated any more. So we are forced to distinguish between the interpretations in the stochastic case.

Turning now to the continuity equation, it should be remembered that the underlying conservation law is an integral law, stating that the solute mass in a closed volume can only change as a result of the total flux across the surface. The continuity equation is merely a differential statement of this based on Taylor expansion of the flux, and assumes that the flux and hence also the velocity can be differentiated. So there is clearly a conceptual problem in applying the continuity equation to the particle velocity in stochastic flow. On the other hand, solute conservation as applied to a single fluid element holds trivially, at least in the absence of diffusion, which is the case we are mostly concerned with.

One way out of this difficulty is to conclude that the conservation law should be applied to the complete flow. By this we mean the superposition of individual realizations of the flow trajectory, just as the deterministic flow field is the superposition of trajectories of all fluid elements. Therefore $\underline{v}(\underline{x})$ in equation (7.5) should be interpreted as the average velocity, averaged over all realizations that pass through position $\underline{x}$. We have seen in section 5.6.3 that when a stochastic population was averaged over all realizations the deterministic value was obtained; below it will be shown that the same holds for the stochastic flow velocity. So in the end we are justified to use the deterministic equation (7.2) to replace the unknown velocity $\underline{v}$ in equation (7.5) by the known, non-stochastic function $\underline{u}(\underline{x})$.

Another way out is to recognize that the continuity equation is not really part of the stochastic problem, and that solute mass conservation should rather be applied in another way after solving the SDE equations. This will be further pursued below, but first we investigate where the use of $\underline{u}(\underline{x})$ in equation (7.5) leads.

The next problem in obtaining an ID from equation (7.5), is that a total time derivative is needed on the left so that it can be written as $dc = b(c)\, dt$, to use the notation of equation (5.3). That is easily addressed by transferring the second term in equation (7.5) to the left and we are left with

$$\frac{dc}{dt} = -c\nabla \cdot u + \nabla D_m \cdot \nabla c + D_m \nabla^2 c. \tag{7.8}$$

The last two terms on the right still do not conform to the ID form, as they involve derivatives of the unknown c. That could be remedied by taking e.g. ∇c and $\nabla^2 c$ as new variables and adding more equations. This leads to an infinite hierarchy of equations, that would have to be truncated at some stage and does not look very promising. Another idea is to solve equation (7.8) iteratively; i.e., put an assumed form for c in the last two terms, solve the equations and put the resulting c back as the next iteration. In particular, we might take $c = 0$ as the first guess; that seems appealing, since it is equivalent to taking $D_m = 0$ and as discussed in chapter 1, microscopic diffusion is indeed found to be negligible in many experimental situations.

In what follows, we do make the simplifying assumption that diffusion can be neglected. However, equation (7.8) provides a clear way in which one could in principle move beyond this assumption and add diffusion effects at a later stage. Taking $D_m = 0$ equation (7.8) reduces finally to the desired form. We could now proceed to combine it with equation (7.2) and set up the problem as a set of two ID equations. That would mean adding Wiener process terms to the right hand sides of both equations. However, the validity of this may be questioned.

From a physical perspective, it is clear that the random variations of the position increments represent the deflections suffered by a fluid element as it meanders thorough the porous medium; but there is no additional physical mechanism independent of this to randomize the concentration. Mathematically, one adds the Wiener terms to represent random variations in the driving coefficients of the original deterministic equation. But in equation (7.8) the driving coefficient is $\nabla . u$, which is not independent of the driving coefficient u in equation (7.1). In fact the point of the eigenfunction sum in equation (7.1) was to explicitly introduce spatial correlation between random variations in u at neighboring points, and adding random variation to the spatial derivative would contradict this.

Therefore it is only appropriate to add the Wiener terms to equation (7.1) while (7.8) should remain as it is, a deterministic equation. Formally, it is possible to carry such a deterministic equation along in a set of ID's, but as there is no coupling between the variables in the equations, this does not contribute anything and creates unnecessary baggage. So, once more, we are led to the conclusion that the continuity equation is not inherently part of the stochastic model and that solute mass conservation should be enforced in a different way.

7.2 Integral Formulation of Solute Mass Conservation

Consider, first, a deterministic flow of an incompressible carrier fluid specified by a stationary velocity field $u(x)>0$, and transporting a solute with concentration $c(x,t)$ without diffusion. We consider a 1-dimensional description, i.e. u and c have been averaged out over the other two dimensions and c represents the solute mass per unit length. Suppose that the solute is introduced into the flow at an initial time t_0 with a localized distribution, e.g $c(x,t_0)$ may be a Gaussian peak. The problem is to find the concentration at a later time t, subject to solute mass conservation.

Because the fluid is incompressible, conservation of the fluid mass requires that the volume of fluid passing a point in an interval Δt is independent of position and time, i.e.

$$A(x)u(x)\Delta t = A(x')u(x')\Delta t, \quad \forall t \tag{7.9}$$

where $A(x)$ is the cross-sectional area perpendicular to the flow.

The total amount of solute that passes x between t and $t+\Delta t$ is given by $c(x,t)$ $u(x)$ Δt , but clearly a conservation equation of the type of equation (7.9) does not apply to this as the amount passing a given point changes with time. Nevertheless, as there is no diffusion, we can still make the statement that all the solute contained in the volume element $A(x')$ $u(x')$ Δt will pass a point $x > x'$ in a time interval of the same length Δt ,at the later time t when that particular volume element reaches x.

Defining the kinematic time interval $w(x|x')$ for traveling from x' to x by

$$w(x \mid x') = \int_{x'}^{x} \frac{dx}{u(x)} \tag{7.10}$$

that statement is expressed by

$$c(x,t)u(x)\Delta t = c(x',t - w(x \mid x'))u(x')\Delta t \,. \tag{7.11}$$

This is equivalent to the following equation, which appears more complicated but is formulated in a way that allows for later generalization:

$$c(x,t) = \frac{u(x_0)}{u(x)} \int_{-\infty}^{t} dt' \, c(x_0,t')\, \delta(t' - t + w(x \mid x_0)) \,. \tag{7.12}$$

This form is appropriate for a boundary value problem, where the value of the concentration is known at the fixed boundary x_0 for all times earlier than t. If instead, as originally stated above, we know the initial concentration at $t = t_0$, we have an initial value problem and to formulate the conservation law we need to first solve for x' from the equation

$$t_0 = t - w(x \mid x') \quad \Rightarrow \quad x' = \chi(x,t \mid t_0) \tag{7.13}$$

and then rewrite equation (7.11) as

$$\begin{aligned} c(x,t) &= \frac{1}{u(x)} \left[c(x',t_0)u(x') \right]\Big|_{x'=\chi}, \\ &= \frac{1}{u(x)} \int_{-\infty}^{\infty} dx' c(x',t_0)u(x')\delta(x' - \chi(x,t \mid t_0)). \end{aligned} \tag{7.14}$$

To illustrate the use of equations (7.12) and (7.14) as conservation laws, consider first the simple case where $u(x)$ is constant; say $u(x) = v_0$. Then we have

$$w(x \mid x') = \frac{x - x'}{v_0}; \qquad \chi(x,t \mid t_0) = x - v_0(t - t_0)\,, \tag{7.15}$$

and this leads respectively for the boundary value and initial value problems to the results

$$\begin{aligned} c(x,t) &= c(x_0, t - \tfrac{x - x_0}{v_0}), \\ c(x,t) &= c(x - v_0(t - t_0), t_0). \end{aligned} \tag{7.16}$$

Both of these have the straightforward interpretation of simple plug flow, i.e. regardless of the shape of the input solute concentration, this is simply translated forward unchanged at the flow speed v_0.

The situation is a little more interesting when u changes with x. The simplest example of this is when we have a linear dependence, which we write as $u(x) = v_0 + p(x-x_0)$. In this case equation (7.10) yields

$$w(x \mid x') = \frac{1}{p} \ln\left(\frac{v_0 + p(x - x_0)}{v_0 + p(x' - x_0)} \right) \tag{7.17}$$

provided that the x-value at which $u(x)$ becomes 0, is excluded from the interval $[x',x]$. If this x-value, referred to as a stagnation point of the flow, is inside the interval, the mathematical treatment might be salvaged by splitting the interval in two. More generally (for flow in the positive x-direction) we can say that the interval of applicability of the expressions to be derived, are limited from below by the stagnation point for accelerating flow, and from above for decelerating flow. The exclusion of stagnation points can also be justified on physical grounds, as is discussed further below.

Using equation (7.17) to calculate χ leads to the expressions

$$\frac{u(x')}{u(x)} = \frac{v_0 + p(x' - x_0)}{v_0 + p(x - x_0)} = e^{-p(t-t')}, \text{ and} \tag{7.18}$$

$$\chi(x,t \mid t_0) = x_0 + \frac{v_0}{p}\left[e^{-p(t-t_0)} - 1 \right] + (x - x_0)\, e^{-p(t-t_0)}. \tag{7.19}$$

It is instructive to take the concrete example where we have an initial value problem, specified by an initial Gaussian concentration peak centered at $x = \xi$ and with a variance s^2:

$$c(x,t_0) = \frac{1}{\sqrt{2\pi s^2}} exp\left[\frac{-(x-\xi)^2}{2s^2} \right]. \tag{7.20}$$

The choice of a normalized Gaussian implies that we have for simplicity scaled the total solute mass to unity. Another useful simplification follows by noting that in the initial value problem there is no fixed boundary, i.e. the value of x_0 just determines the value of v_0 and we can just as well choose $x_0 = \xi$.

Equations (7.20) and (7.19) can now be substituted into (7.14). The resulting expression is simplified by noting that from (7.18)

$$\frac{u(\chi)}{u(x)} = e^{-p(t-t_0)} \tag{7.21}$$

and that with our choice of x_0, the expression $(\chi\text{-}\xi)$ reduces to

$$\chi - \xi = e^{-p(t-t_0)}(x - X(t)), \text{ and} \tag{7.22}$$

$$X(t) = \xi + \frac{v_0}{p}\left[e^{p(t-t_0)} - 1\right]. \tag{7.23}$$

The expression X(t) has a straightforward physical interpretation. Going back to equation (7.13), note that the first equation merely represents the relationship between the initial position x' (at time t_0) or *source point* of the fluid element that is found at the *target point* x at a later time t . So χ is the function that calculates the source point, given a target point. Conversely, $X(t)$ is the target point that originates from the particular source point $x' = \xi$ (as is easily confirmed by putting $\chi = \xi$ in equation (7.19)). In other words, X(t) represents the trajectory followed by the fluid element that contained the peak value of the concentration at the initial time.

We can now put together equations (7.14) and (7.20) to (7.23) to obtain the evolution of the concentration in a linearly accelerating flow, without diffusion or dispersion, as

$$c(x,t) = \frac{1}{\sqrt{2\pi\,\sigma(t)^2}}\,exp\left[\frac{-(x - X(t))^2}{2\sigma(t)^2}\right] \tag{7.24}$$

where

$$\sigma(t) = s\exp[p(t - t_0)]\ . \tag{7.25}$$

Equation (7.24) demonstrates some striking features:

- a concentration peak that starts off as a Gaussian, retains its Gaussian shape while propagating;
- the Gaussian remains normalized, i.e. the total solute mass is conserved;
- the peak of the Gaussian moves at the speed predicted by the kinematics of the fluid flow;

- the width of the Gaussian grows exponentially in an accelerating flow ($p > 0$) and shrinks exponentially in a decelerating flow ($p < 0$); and
- for $p = 0$ the plug flow solution is regained.

The change in the extension of the concentration plume is easily understood qualitatively. In an accelerating flow, parts of the plume ahead of the peak value are always moving faster than those at the peak, and parts behind the peak move slower than the peak so that the plume is stretched out by the acceleration; and conversely it is compressed for deceleration. This effect appears superficially like dispersion, but is merely a result of the kinematics of the fluid motion. Unlike "real" dispersion, it is fully reversible in the sense that if, after propagating for a time Δt the acceleration is instantaneously reversed, equation (7.25) implies that the Gaussian plume will return to its original extension after a further time interval Δt. We will refer to this effect as *kinematical dispersion* in what follows.

However, the surprising thing is that this does not distort the peak shape. This property is a peculiarity of the linear acceleration; it depends on the fact that the factor $u(\chi)/u(x)$ required by solute conservation turned out in equation (7.21) to depend on time only, and that is a consequence of the logarithm in the expression for $w(x|x')$ in equation (7.17). For any other dependence of the velocity on position, distortions of the Gaussian shape are to be expected. That concludes our discussion of the integral form of the conservation law for deterministic solute transport, and we turn to the generalization of equations (7.12) and (7.14) to the stochastic transport model.

The latter case, that of an initial value problem, is quite straightforward. The input to this problem is a given initial spatial concentration profile, $c(x,t_0)$. Equation (7.14) can be interpreted in a physical way as saying that we divide up this concentration into fluid elements, one for each value of x'. For each of these, the position x at which it will end up at time t, is exactly determined from the flow velocity field. Conversely, to determine the concentration at x at time t, we merely need to identify the source point from which the target point fluid element originated, and that is exactly the effect of the Dirac δ-function in the integral of equation (7.14). It picks out a single fluid element from the original concentration profile, and if there has been a velocity change while the fluid element moved from the source point to the target point, the concentration is corrected by the ratio of the velocities as required by solute conservation. Obviously, then, in a stochastic model we take into account that there is not a one-to-one correspondence between a source point and a target point any more, but rather for any fluid element in the original concentration there is a probability that it will reach the chosen target point. So we replace the certainty of motion from the source point to the target point that the δ-function expresses, by a probability density for such motion. Then the generalization of equation (7.14) is

$$c(x,t)=\frac{1}{u(x)}\int_{-\infty}^{\infty}dx'c(x',t_0)u(x')P_{t_0}(x'\mid x,t)) \tag{7.26}$$

where we define $P_{t'}(x'|x,t)$ as the probability density over the position variable x', that a fluid element that is known to arrive at the target point (x,t), has originated from x' at a time t'.

In a similar vein, the boundary value conservation equation might be written

$$c(x,t)=\frac{u(x_0)}{u(x)}\int_{-\infty}^{t}dt'c(x_0,t')P_{x_0}(t'\mid x,t)) \tag{7.27}$$

where $P_{x'}(t'|x,t)$ is the probability density over time t', that a fluid element that is known to arrive at the target point (x,t), has originated from x' at a time t'. The two probability densities are not the same, but are related through an appropriate variable transformation.

While the initial value and boundary value formulations appear quite similar, as indeed they are in the deterministic case, there are subtle differences in how they can be applied in the stochastic situation. These differences are highlighted by considering how the input concentration might be prepared physically. In the case of the initial value problem, at least at the level of a gedanken experiment, there is no conceptual difficulty in dropping a previously prepared spatial concentration profile instantaneously into the carrier fluid flow at the instant $t = t_0$.

For a boundary value problem, the physical realization would be to inject solute into the flow at the point $x = x_0$ with a predetermined time profile, say $f(t)$. However, the difficulty is that as a moment's consideration shows, the concentration profile in the actual flow at point x_0 , $c(x_0,t)$, will be different from $f(t)$. That is because the stochasticity implies that some of the solute concentration will be carried upstream of the injection point. For example, if $f(t)$ is only a momentary pulse at $t = 0$, we would have $c(x_0,0+) = ½f(0+)$ because over a short enough time interval the translation of the carrier fluid can be neglected and equal amounts of solute would be taken upstream and downstream by stochastic fluid displacements. Subsequently, some of the solute that was taken upstream by stochastic displacements will be carried past x_0 by the flow, but a diminishing fraction will always remain upstream.

The net effect is that any abrupt change in the injection profile will be rounded off in the time profile at the injection point inside the flow.

The implication is that while a sharp edged spatial concentration profile (for example, a square pulse) can be introduced as an initial value, it is not possible to produce such a spatial profile by injection at a boundary point in the case of stochastic flow because of the inevitable rounding effect. That is in contrast to deterministic flow, where the initial value and boundary formulations are equivalent in the sense that any state specified in an initial value problem, could be prepared by an appropriate injection time profile at a suitably chosen boundary point.

A related remark stems from the observation that any initial value concentration of interest would be localized in space. That places limitations on the specification of a related boundary value problem, because of the fact that stochasticity leads to spreading. Consequently, an injection point cannot be put too far away upstream from the locality of the initial concentration, otherwise even the sharpest injection peak would have spread out too much by the time that it reaches that location. By the same token, for a chosen boundary point, there is an earliest possible time that the injection can start in order to produce a concentration with the appropriate degree of localization at the time t_0 that specifies the initial value version of the problem. In other words, a cutoff time needs to be introduced into equation (7.27) instead of the $-\infty$ integration limit, for a boundary value problem to represent a localized concentration in a related initial value formulation.

These somewhat vague remarks will have to be further explored and refined when we have obtained more quantitative information e.g. on the rate at which stochastic spreading takes place. For the moment, however, they serve as warning about the dangers of applying intuitive ideas based on deterministic models, directly to the stochastic model. To avoid such difficulties for now, we will base our analysis below on the conceptually simpler initial value formulation represented by equation (7.26) even though the boundary value version is closer to the physical situation in actual experiments.

The main conclusion reached here is that in order to calculate the concentration in an integral formulation of solute mass conservation, we need to find the probability distribution of fluid elements that result from the stochastic model. That task is addressed in the next section.

7.3 Stochastic Transport in a Constant Flow Velocity

After considerable preparation, we are now finally in a position to apply the SDE theory based on Dynkin's equation to solute transport modeling. Having disposed in the previous sections of the problem how to include solute mass conservation into the model, we are left with the single ID equation given by equation (7.1) to process further. This is already in the correct form according to equation (4.3) provided that we interpret dB_t in the latter as a vector, with components given by $db_n(\omega,t)$. To avoid complicating the argument by having to deal with a vector in an infinite dimensional space, we take the pragmatic view that the eigenfunction expansion can be truncated after M terms. In this case σ in equation (5.3) is also an M-dimensional vector, and by comparison with equation (7.1) it is given by

$$\sigma(x) = \left(\sqrt{\lambda_0}\, f_0(x), \sqrt{\lambda_1}\, f_1(x), \ldots, \sqrt{\lambda_M}\, f_M(x)\right) \tag{7.28}$$

whereas the drift coefficient in equation (7.1) is simply the scalar function $u(x)$. The next step is to construct the generator of the ID according to equation (4.11). For this we need

$$\sigma \cdot \sigma^T = \sum_{n=0}^{M} \lambda_n f_n(x) f_n(x) = C(x,x) \tag{7.29}$$

which is also a scalar function, and for the second equality we have used equation (6.6). Note that in the final result M does not appear any more, so that the expansion can be truncated at any large finite M without needing to explicitly solve the eigenvalue problem.

In fact, we see from equation (6.11) that for the covariance kernel that was proposed in Chapter 6, $C(x,x)$ reduces to a constant (1), if we assume that the amplitude factor in equation (6.11) has been absorbed into γ^2 as explained in the introduction). This reflects the assumption that the covariance of realizations of the α-processes of equation (6.4) at position x, is the same for all positions. This seems reasonable enough, although it is possible to construct a generalization where the time-dependent Wiener processes are not only generalized Wiener processes instead of standard ones, but in addition have different q-functions (in the notation of equation (2.10)) at different positions. In this case the spatial correlation kernel can be such that $C(x,x)$ is a function of position. However, as there does not seem to be any physical reason to assume this more complicated behavior, we restrict ourselves to the case that $C(x,x) = 1$. Then, we can write down the generator equation as

$$A\,f(x,t) = u(x)\frac{\partial f}{\partial x} + \frac{1}{2}\gamma^2\frac{\partial^2 f}{\partial x^2}. \tag{7.30}$$

Bearing in mind that the generator is used to calculate expectation values and probabilities as detailed below, the fact that equation (7.30) does not contain any reference to the spatial correlation kernel any more, is very significant. It means that while individual realizations of the motion of a fluid element are affected by the functional form of the kernel, all such dependence is averaged out when taking expectation values over all representations. This is not obvious, and might be quite hard to determine from numerical simulations of individual realizations. The ease with which this conclusion follows from the potential theory approach is a striking illustration of its power.

As in the population growth example, different questions about the flow can be answered by solving the cases $A\,f = 0$, $A\,f = K$ and $A\,f = \partial f/\partial t$. The first two of these relate respectively to questions about the probability that a fluid element starting from the origin reaches a predetermined position, and the expected time for it to do so. However, as the results agree with what would be expected from a deterministic model we do not dwell on them.

The third case on the other hand, is the one that eventually yields the probability distribution that is needed to implement solute mass conservation. In this section we limit ourselves to the case that $u(x)$ is a constant, say v_0. That would for example apply in the case of a constant pressure gradient across a homogeneous porous medium described by the Darcy equation, where we would have $v_0 = -(K/\varphi)(d\phi/dx)$. To facilitate solution, we transform the generator equation to a new position variable z and time variable T defined as

$$z = \frac{x\,v_0}{\gamma^2}\;;\quad T = \frac{t\,v_0^{\,2}}{\gamma^2}. \tag{7.31}$$

These are, in some sense, also scaling transformations to “dimensionless” parameters, but are distinct from the physical scaling that was mentioned in the introduction of this chapter. Here, we are merely dealing with a mathematical transformation that scales the various terms of the differential equation similarly, i.e. to a form in which all coefficients are simple numbers of order unity:

$$\frac{1}{2}\frac{\partial^2 f}{\partial z^2}+\frac{\partial f}{\partial z}=\frac{\partial f}{\partial T}. \tag{7.32}$$

In the standard way this partial differential equation is reduced to an ordinary one by separating variables in writing

$$f(z,T)=e^{-KT}f_K(z) \tag{7.33}$$

where K is introduced as a separation constant to be determined. It is easily seen that the resulting equation for f_K is solved by a simple exponential form, $f_K(z) = exp(-\alpha z)$, provided that α is related to the separation constant by

$$K=\alpha-\tfrac{1}{2}\alpha^2. \tag{7.34}$$

Finally, we revert back to the coordinate pair (x,t). The resulting expression is cast in a particularly simple form by making use of the freedom allowed by the fact that K and therefore also α are arbitrary constants. Hence we can as well write α in terms of a new arbitrary constant a which we define by $\alpha = a\gamma^2 / v_0$ to obtain the solution in the form:

$$f(x,t)=e^{-ax}\,e^{a v_0 t-\frac{1}{2}a^2\gamma^2 t}. \tag{7.35}$$

What we have achieved here, is to find a solution of equation (5.14) for the case of our solute transport model ID – in fact, due to the arbitrary nature of a, we have found not just one but a whole family of solutions. We can now invoke equations (5.13) and (5.15) to write down Kolmogorov's version of the Dynkin equation, just as was done for the population growth model:

$$E^0\left[e^{-aX_t}\right]=e^{a v_0 t-\frac{1}{2}a^2\gamma^2 t}. \tag{7.36}$$

To interpret this equation, it is useful to restate it in terms of an operational definition. We start with the position x of a fluid element at a time t. From this, we calculate the function e^{-ax}. Rather than x just being an ordinary function of t, we allow x to vary stochastically as represented by the stochastic process X_t. Then the expectation value at $t > 0$ of the function, taken over all realizations of the process that start from the initial value $x = 0$ at $t = 0$, is given by the right hand side of equation (7.36).

This expectation value can be expressed in an alternative way by using a probability density as introduced in the previous section. Let $P_t(x|x',t')$ be the x-domain probability density that a fluid element known to start from x' at a time t', arrives at the target point x at the time $t > t'$. Then we have

$$E^0\left[e^{-aX_t}\right] = \int_{-\infty}^{\infty} dx\, e^{-ax} P_t(x\,|\,0,0). \tag{7.37}$$

This equation is analogous to equation (5.21), except that in this case there is an infinity of possible values of e^{-ax} at the stopping time, instead of just two possible values as in the former case.

Combining equations (7.36) and (7.37) we now have an integral equation to solve for the probability density:

$$\int_{-\infty}^{\infty} dx\, e^{-ax} P_t(x\,|\,0,0) = e^{a v_0 t - \frac{1}{2} a^2 \gamma^2 t}. \tag{7.38}$$

As remarked at the end of Chapter 5, a formal approach might be taken to solve it. However, we present a more intuitive approach. The first step is to notice that if we make an expansion in powers of a on both sides of the integral equation, coefficients may be set equal because a is an arbitrary constant. This leads to expressions for all moments of the probability density and in particular to the following expressions for the mean and variance of the position:

$$\begin{aligned} &\langle x \rangle = v_0 t; \\ &Var(x) = \langle x^2 \rangle - \langle x \rangle^2 = \gamma^2 t. \end{aligned} \tag{7.39}$$

Also, by putting t = 0 in equation (7.38) it is seen that as the equation holds for any value of a, we must have $P_0(x|0,0) = \delta(x)$. So the solution of the integral equation should be a function that starts at $t = 0$ as a delta-function peak at the origin, then develops as a peak centered near $v_0 t$ and with a variance increasing proportional to t. An obvious guess with these properties is a Gaussian peak of the form

$$P_t(x\,|\,0,0) = \frac{1}{\sqrt{2\pi\gamma^2 t}} \exp\left[\frac{-(x - v_0 t)^2}{2\gamma^2 t}\right]. \tag{7.40}$$

Indeed, if we substitute equation (7.40) into (7.38), complete squares in the exponent and then perform the integration by use of the standard result

$$\int_{-\infty}^{\infty} e^{-y^2} dy = \sqrt{\pi} \ , \tag{7.41}$$

it is seen that equation (7.40) is an exact solution. If the time and space origins are shifted to a different position, it is clear that the probability distribution can still only depend on the difference between the source and target point coordinates, i.e.

$$P_t(x \mid x', t') = \frac{1}{\sqrt{2\pi\gamma^2 (t-t')}} \exp\left[\frac{-((x-x') - v_0(t-t'))^2}{2\gamma^2 (t-t')} \right]. \tag{7.42}$$

The probability density required for calculating the solute concentration according to equation (7.26) is similar to that of equation (7.42), but not identical in principle. For the former, it is the target point *(x,t)* that is fixed and we specify the probability for a source point *(x',t')*; but in equation (7.42) it is the other way around. In the case of a constant flow velocity that we discuss here, a symmetry argument could be made out to show that their functional forms are the same. However, as this result does not hold for more complicated situations, we prefer to derive the relation between the two by a more elaborate argument that can be generalized for example to the case of linearly accelerating flow treated in the next section. This based on noting that, since a fluid element known to be at the source point *x'* at *t'* must end up somewhere at a time *t*, we have

$$\int_{-\infty}^{\infty} dx\, P_t(x \mid x', t') = 1 \, . \tag{7.43}$$

This equation states that when the function *P* of the four variables *(x,t,x',t')* is integrated out over only one of them, namely *x*, all three of the others also drop out of the result. That can only happen if the function is such that it is possible to find a transformation to a new variable *y = y(x,t,x',t')* in terms of which

$$\int_{y_{-\infty}}^{y_{\infty}} dy\, P(y) = 1 \tag{7.44}$$

where $y_{\pm\infty} = f(\pm\infty, t, x', t')$. In other words, the single variable y connects the point *(x',t')* with another point *(x,t)* with $t > t'$, and P(y) gives the probability that these points are related as source and target points respectively, as a density in the combined variable. To retrieve $P_t(x|x',t')$ one would transform the integration variable in equation (7.44) back to x, and by the same token $P_{t'}(x'|x,t)$ is obtained if we transform the integration variable to x'. So

$$\begin{aligned} P_t(x \mid x',t') &= P\left[y(x,t,x',t')\right]\frac{\partial y}{\partial x}, \\ P_{t'}(x' \mid x,t) &= P\left[y(x,t,x',t')\right]\frac{\partial y}{\partial x'}. \end{aligned} \tag{7.45}$$

Time domain probability densities as needed for the boundary value formulation of solute conservation in equation (7.27), can obviously be obtained in a similar way. Applying these ideas to equation (7.42) we can identify

$$y(x,t,x',t') = \frac{(x-x') - v_0(t-t')}{\sqrt{2\gamma^2 (t-t')}}. \tag{7.46}$$

Substituting equation (7.46) into (7.45), we only get a sign change in $P_{t'}(x'|x,t)$ compared to $P_t(x|x',t')$, but that is cancelled by the exchange of the integration limits needed in the former case. So in the end, the right hand side of equation (7.42) represents both of these probability densities for constant flow velocity. In the next section we will see that that is not true any more in accelerating flow.

We are now in a position to calculate the concentration from equation (7.26) given any initial concentration. The simplest case is when we only have a point source of solute that adds a unit mass of solute to the flow at the point $x = \xi$, i.e. we take $c(x,t_0) = \delta(x-\xi)$. As $u(x) = u(x') = v_0$ it is trivial to see that this leads to

$$c(x,t) = \frac{1}{\sqrt{2\pi\gamma^2 (t-t_0)}} \exp\left[\frac{-((x-\xi) - v_0(t-t_0))^2}{2\gamma^2 (t-t_0)}\right]. \tag{7.47}$$

In other words, the delta-function initial concentration peak develops into a Gaussian peak that moves at the constant speed v_0, and spreads, with a variance that increases linearly with time, as illustrated in Figure 7.1.

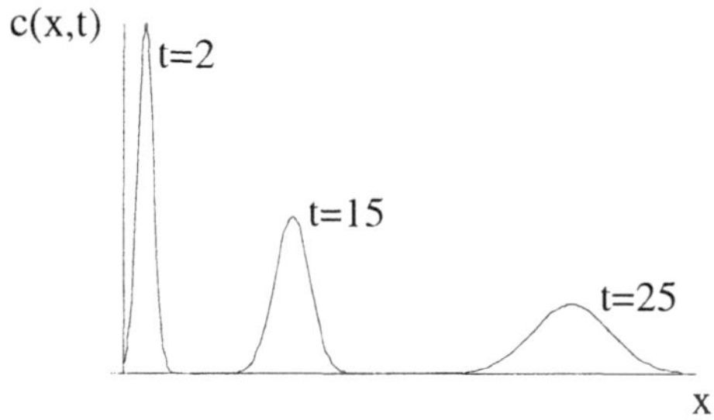

Figure 7.1 Development of concentration peaks.

The curves in the figure represent "snapshots" of the concentration profile at three consecutive times, with a positive value for v_0, i.e. the flow direction is to the right.

Moreover, we can conclude that if the initial concentration was a Gaussian peak rather than a delta function, it would also develop as a Gaussian that is simultaneously translated at the flow velocity and spreads proportional to time. This follows logically since if we by divide the time interval (t_0,t) into two at an arbitrary intermediate point t_1 , the concentration at t_1 will be a Gaussian according to the result just derived; and its subsequent development cannot depend on whether the concentration at t_1 was physically prepared by directly introducing a Gaussian distribution into the flow at t_1 or by inserting a point source at the earlier time t_0. Of course, the same result can be proved in a more tedious fashion mathematically by inserting a Gaussian initial concentration into equation (7.26) and performing the integral by completing squares in the exponent.

Before proceeding to make direct connection to physical observations, we need to remember that the variables in equation (7.47) are still scaled variables. The scaling constants ℓ_x and ℓ_t are to be chosen to reflect the physical mechanisms that cause stochasticity. In our approach, stochasticity is used to model the deflections of fluid flow by pore walls. Individual displacements suffered by a fluid element when it hits a pore wall has to be on the scale of a pore diameter, and time intervals between deflections on the scale of the time it needs to traverse a pore. Therefore, it is plausible that ℓ_x should be of the order of a typical pore diameter, and that ℓ_t is chosen as ℓ_x/v_0. Having already introduced the correlation length b in equation (6.10) as a pore scale constant, it is consistent to use that also here for the length scale. That leads to the conclusion that if equation (7.47) is rewritten in terms of ordinary unscaled laboratory coordinates, it takes the form

$$c(x,t)=\frac{1}{\sqrt{2\pi\,\gamma^2 b\,v_0\,(t-t_0)}}\exp\left[\frac{-\left((x-\xi)-v_0(t-t_0)\right)^2}{2\gamma^2\,b\,v_0\,(t-t_0)}\right]. \tag{7.48}$$

The variance of this concentration, about its uniformly translating peak value, is given by

$$\sigma^2(t)=\gamma^2\,b\,v_0(t-t_0). \tag{7.49}$$

The standard deviation, σ, is a natural measure of the extension of a concentration plume and we see that this increases proportional to the square root of the time. That is a well-known behavior usually associated with the advection-dispersion equation (ADE), and so it is useful to establish the connection. We write the ADE (1.9) in the following form, where the porosity has been absorbed into the concentration and a constant flow velocity is assumed:

$$\frac{\partial c}{\partial t}+v_0\frac{\partial c}{\partial x}-D\frac{\partial^2 c}{\partial x^2}=0. \tag{7.50}$$

Here, D represents the coefficient of hydrodynamic dispersion. The form of this equation closely resembles that of the time-dependent generator equation (7.32) but has a different sign for the spatial first derivative term. It may be confirmed by direct substitution that it is solved by a translating, dilating Gaussian of the form $\exp\left(\frac{-(x-v_0 t)^2}{2\sigma^2}\right)$ provided that

$$\sigma^2(t)=(t-t_0)\,2D. \tag{7.51}$$

In other words, the ADE equation also predicts that the variance of a Gaussian concentration peak will grow linearly with time and identifies the coefficient of this growth as twice the dispersion coefficient. Some interesting conclusions can be drawn from the comparison:

- the microscopic stochastic model agrees with the macroscopic ADE equation for the case of a constant flow velocity; and
- by comparing equations (7.49) and (7.51) we can write the macroscopic, empirically defined dispersion coefficient in terms of the parameters of the microscopic model:

$$D = \tfrac{1}{2}\gamma^2 b\, v_0 . \qquad (7.52)$$

The fact that the dispersion coefficient is proportional to the flow velocity follows in a natural way from the stochastic model. This is a crucial assumption in the commonly used definition of dispersivity as an intrinsic materials property of a porous medium – see section 1.3.3. Moreover, the dispersivity is also obtained in a natural way as a length on the scale of the pore diameter, again in agreement with work based on the ADE equation and shown in Table 1.1.

In the case of the ADE equation, the dispersion coefficient D is the fundamental empirical parameter and it is a consequence of the equation that a Gaussian concentration plume remains Gaussian but spreads in a way characterized by D. In the stochastic model, there is no dispersion coefficient as such but we can determine an effective value for it by reversing the argument. We start with a Gaussian plume, calculate its time evolution and then extract a coefficient for the linear growth of the variance with time.

That is only really plausible in a case like the one we discussed here where the variance of the Gaussian does grow in a linear fashion. Any other time dependence would reduce this procedure at best to an approximate one, where we might determine an effective D value, but which changes with time. At a more fundamental level, however, one could conclude from such a situation that the concept of dispersion as a phenomenon that is similar to diffusion, only on a larger scale, is flawed.

From this remark it is clear that although a study of the time evolution of a Gaussian plume may appear to be a very narrow focus on a special case, an understanding of it can contribute significantly to insights into the broader phenomenon.

The idea of replacing the definition of the dispersion constant as a Fickian coefficient, by one based on the variance of the concentration plume is one that is commonly used. In experimental work, the variance is a well-defined global property of the plume that is more easily measured than the local concentration gradient and flux needed for the Fickian definition. This idea is applied for example in the work of Moroni and Cushman (2001), where the assumption of a linear time dependence is avoided by defining the dispersion coefficient as half the first time derivative of the variance.

If the Fickian description of dispersion contained in the ADE equation holds, the two definitions of the dispersion coefficient are consistent. The work in this section shows that the stochastic model supports such consistency in the case of a constant flow velocity.

However, more generally it does not necessarily follow that a dispersion coefficient obtained from variance measurements can consistently be used in the ADE equation. One would first have to show that the more detailed predictions of the ADE equation, such as the retention of an initially Gaussian concentration profile, but with a linear time dependence of the variance, also hold. For example, one might hypothetically picture a more complicated relation between the solute flux and the concentration gradient that causes the shape of the profile to change away from a Gaussian one but leaving the variance unchanged. Then the variance based or "global" value of the dispersion coefficient would be zero while on a local scale there is still dispersion manifested by solute flux in response to a concentration gradient.

The discussion above raises some interesting philosophical points about the relation of the stochastic dispersion model to diffusion. As explained after equation (7.8), in setting up the model that we solved above, ordinary physical (i.e. molecular scale) diffusion has been left out deliberately – but the final result, that a momentary point source develops into a spreading Gaussian, is in fact exactly what one expects from ordinary diffusion. From a mathematical perspective, this is not surprising; diffusion is after all the prototypical SDE pheneomenon and the methods we used, based on ID's and Dynkin's equation, are in fact mathematical generalizations of diffusion theory. From a physical perspective, one might say that molecular diffusion results from stochastic displacements of individual solute molecules, as a result of molecular collisions (Brownian motion), and this is indeed very similar to our dispersion model where we have displacements of macroscopic fluid elements as a result of collisions with pore walls.

There are two striking differences, though. Firstly, the spatial scale is much larger, leading to a dispersion coefficient that can be orders of magnitude larger than the diffusion coefficient. And secondly, in dispersion the time scale (i.e., the interval between collisions) is determined by the average fluid flow velocity, while in diffusion the time scale is determined by the fluid density, temperature, etc.

For a constant flow velocity v the close connection between dispersion and the velocity only manifests in the fact that the dispersion constant is proportional to v, and we might describe this as a diffusion-like or "diffusive" dispersion behavior. There are more profound consequences for dispersion when v changes. That is the subject of the next section.

7.4 Stochastic Transport in a Flow with a Velocity Gradient

Applying the same model as before to study dispersion in a flow with a variable flow velocity, we need to go back to equation (7.30) and would ideally like to solve the generator equation for an undetermined $u(x)$. Unfortunately it is not possible to find an analytic solution to the generator equation in this general case. Of course, for any particular $u(x)$ determined from some appropriate model of the fluid flow, it would in principle be possible to find a solution to the generator equation numerically. However, we saw in the previous section that the solution of the generator equation only yields an auxiliary function, from which the probability density needs to be extracted by solving an integral equation that is constructed using the auxiliary function. Following this route, the integral equation would also have to be solved numerically; and then finally, the concentration calculated by numerical integration of the solute conservation integral. Apart from possible numerical pitfalls along the way, this approach does not appear likely to yield any insight into the mechanisms of dispersion in a variable flow. For this reason, we restrict ourselves instead to a case where $u(x)$ is simple enough to still allow all three of the listed steps to be done analytically, as was done for the constant flow. We assume that the flow velocity has a straight line dependence on position:

$$u(x) = v_0 + p(x - x_0) = v_0 \pm \mu^2 (x - x_0) \qquad (7.53)$$

where either p or μ^2 represents the velocity gradient, and the second form just facilitates some intermediate steps where it turns out that the algebra depends on whether the flow accelerates or decelerates.

The applicability of this simple assumption may be extended if one considers that any arbitrary variation of u may be approximated by a sequence of x-intervals with a different constant velocity gradient in each, i.e. a piecewise linear approximation for $u(x)$. While that strengthens the case for tackling the constant gradient case, a piecewise approximation is not as simple as it sounds, because one needs to use the boundary value formulation of the conservation equation and as alluded to in the discussion of equation (7.27) complications arise from that. In fact, if one is prepared to deal with those complications, an even simpler model of a variable flow velocity would be a piecewise constant velocity.

Before turning to the mathematics of the solution, it may be helpful to find some concrete examples of a varying flow velocity to serve as mental pictures of the situations that we are trying to describe. A first example would be that of flow along a pipe with a constriction, pictured as a cross-section in Figure 7.2 below:

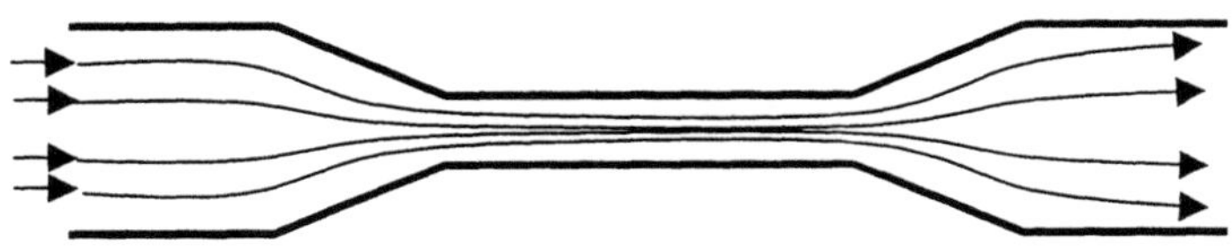

Figure 7.2 Flow along a pipe with a constriction.

The effective 1-dimensional flow velocity along the horizontal axis can be plotted schematically as follows:

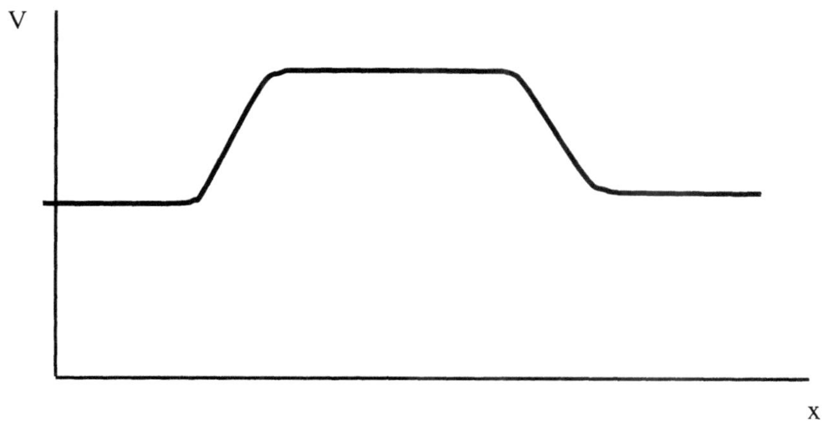

Figure 7.3 Flow velocity along the constricted pipe shown in Figure 7.2.

This velocity profile can obviously be modeled well as piecewise linear, with a constant velocity gradient over the interval representing the conical pipe section. Clearly, the drift velocity profile would be similar if the constricted pipe was filled with a homogeneous porous medium. If the constriction was formed not by the geometrical shape of the pipe, but by an equivalently shaped region of zero-conductivity porous medium inside it, the velocity profile would obviously be the same; and on the other hand, a similar drift velocity profile would also result if the pipe had a constant cross section but the porous medium filling it has a reduced porosity in the central region. These are just simple hypothetical examples but they demonstrate that variations in either the hydraulic conductivity or porosity of a porous medium, such as would be expected to be present in naturally occurring aquifers, are bound to cause local variations in the drift velocity, and that it is not unreasonable to model these as linear in the spatial coordinate (at least as a first approximation).

In a similar vein, we can consider the physical meaning of excluding stagnation points of the flow from our model. Figure 7.4 illustrates how stagnation points might arise physically. A decelerating and accelerating flow velocity region that includes a stagnation point is shown at the top of the figure, and below that a 2-dimensional flow pattern that would give rise to such velocity profile when projected to 1-dimensional flow. The crucial point is that that the sign change (i.e., direction change) of the flow at a stagnation point implies either a source or a sink of the 1-dimensional flow. Excluding these is quite plausible for describing the main problem of solute transport in an underground aquifer. The description of dispersion around a stagnation point is left as a special case that will have to be addressed separately.

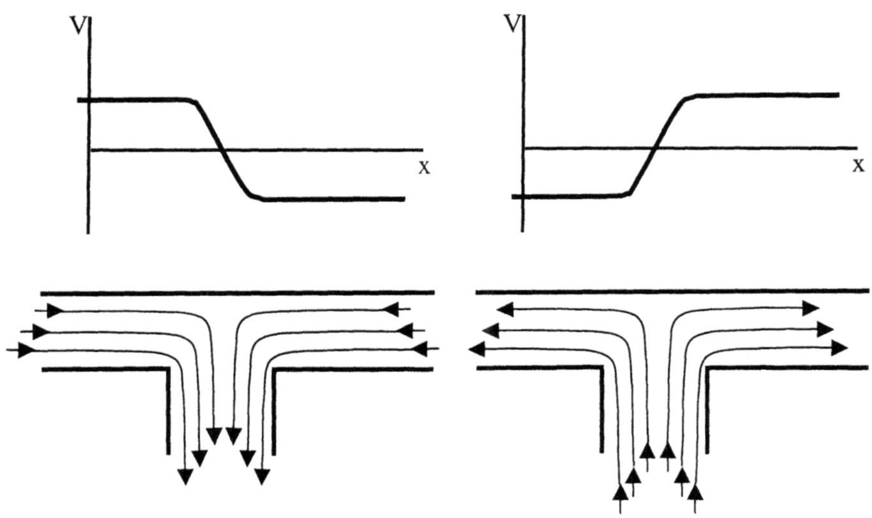

Figure 7.4 Physical stagnation points.

Returning to the mathematical problem, we put equation (7.53) into the generator equation (7.30) and once more transform to scaled position and time coordinates; this time the appropriate definitions are

$$z = \frac{1}{\mu\gamma}\left(\pm\mu^2 (x - x_0) + v_0\right); \; T = \mu^2 t. \tag{7.54}$$

and in terms of these the generator equation is

$$\frac{1}{2}\frac{\partial^2 f}{\partial z^2} \pm z\frac{\partial f}{\partial z} = \frac{\partial f}{\partial t}. \tag{7.55}$$

7.5 Standard Solution of the Generator Equation

The same substitution as before, equation (7.33), is used to separate variables and leads to the following equation for f_K , the spatial part of f:

$$\frac{1}{2}\frac{d^2 f_K}{dz^2} \pm z\frac{df_K}{dz} + K\, f_K(z) = 0. \tag{7.56}$$

Equation (7.56) may be recognized as an example of a standard differential equation, the hypergeometric equation (see Morse and Feshbach, 1953) and it has a pair of independent solutions that can be expressed in terms of confluent hypergeometric functions or Kummer functions, *M(a,b,z)* as follows :

$$f_K^{\pm}(z) = \begin{cases} M(\pm\frac{1}{2}K, \frac{1}{2}, \mp z^2), \\ z\,M(\frac{1}{2}\pm\frac{1}{2}K, \frac{3}{2}, \mp z^2). \end{cases} \tag{7.57}$$

From a compendium of special functions such as that by Abramowitz and Stegun (1965), Table 13.6, we find that Kummer functions with the special values $^1/_2$ and $^3/_2$ for the second argument in fact reduce to Hermite polynomials if the first argument is a negative integer. As K is still an arbitrary separation constant, we can choose values for it to satisfy this requirement.

Considering first the decelerating flow (lower sign) case, it is seen that any even positive integer value of K will be appropriate for the first solution in equation (7.57), and odd integers for the second. In this way both sets of solutions may be collected together and written as

$$f_n^-(z,T) = e^{-nT} e^{\frac{1}{2}z^2} h_n(z), \qquad n = 0,1,2,3... \tag{7.58}$$

where we have introduced the orthonormal functions $h_n(z)$ defined in terms of the Hermite polynomials $H_n(z)$ by

$$h_n(z) = \left(2^n n! \sqrt{\pi}\right)^{-\frac{1}{2}} e^{-\frac{1}{2}z^2} H_n(z). \tag{7.59}$$

The $h_n(z)$ are sometimes referred to as Weber or parabolic cylinder functions and are orthonormal on the z-interval $[-\infty,\infty]$. They form a complete basis, which is expressed mathematically by the closure relation

$$\sum_n h_n(z) h_n(y) = \delta(z-y). \tag{7.60}$$

For the case of accelerating flow (the upper sign in equation (7.57)) a similar set of manipulations can be done but in this case it is first necessary to extract a factor $exp(-z^2)$ by applying the Kummer transformation $M(a,b,z) = e^z M(b-a,b,-z)$ in order to ensure that a real argument for the Hermite polynomials is obtained. The final result is

$$f_n^+(z,T) = e^{-(n+1)T} e^{-\frac{1}{2}z^2} h_n(z), \qquad n = 0,1,2,3... \tag{7.61}$$

At this point we have solved the generator equation and arrived at the equivalent of equation (7.35) in the previous section. Following the same logic invoking Kolmogorov's version of the Dynkin equation, we are now led to the following integral equation corresponding to equation (7.38):

$$\int_{-\infty}^{\infty} dx\, f_n^+(z,0) P_T^+(z \mid z',0) = e^{-\frac{1}{2}z'^2} h_n(z') e^{-(n+1)T} \tag{7.62}$$

where z' is the scaled starting position of the fluid element. This integral equation is easy to solve by multiplying both sides by $h_n(y)$, summing over all n and applying equation (7.60). One finds that

$$\begin{aligned} P_T^+(z \mid z',0) &= e^{\frac{1}{2}(z^2 - z'^2)} \sum_n e^{-(n+1)T} h_n(z') h_n(z), \\ P_T^-(z \mid z',0) &= e^{-\frac{1}{2}(z^2 - z'^2)} \sum_n e^{-nT} h_n(z') h_n(z). \end{aligned} \tag{7.63}$$

Having found fully specified expressions for the probability density for both accelerating and decelerating flows, it might be expected that it is a routine matter to use these in an equation like equation (7.26) to calculate the evolution of an input concentration profile. Unfortunately, it turns out that the solution found is not very useful for practical calculations. One problem with equation (7.63) is that it is hard to see any connection between either of those formulas and the Gaussian expression found for the constant flow probability density in equation (7.42); and after all, either of them must reduce to the constant flow case as $\mu \to 0$. More seriously, we run into convergence problems even for a simple calculation of the moments of the probability distribution. For the case of P^- i.e. deceleration, moments can be calculated, but for P^+ the additional factor $exp(z^2)$ gives rise to a divergent integral inside the summation over n. Obviously the final result cannot diverge, so the problem is that the summation has to be carried out before the integration. Unfortunately however standard mathematical tables do not list a formula for a sum over products of Hermite polynomials as it appears in equation (7.63), nor are symbolic algebra computer packages of any help, suggesting that a calculation of this sum has not been performed before.

Finally, even if we abandon the idea of proceeding with analytical calculations, even for numerical work expansions in terms of Hermite polynomials, which is what equation (7.63) is, can be very slow to converge. This applies in particular to our case, since in the limit as $T \to 0$ we know that the probability density must reduce to a Dirac delta function. The fact that it does so is manifestly clear from equation (7.63) by application of the closure equation (7.60), but numerically plotting the left hand side of the closure relation soon convinces one that truncation of the series at any reasonable value of n gives a very poor representation of the sharpness of a delta function peak.

7.6 Alternate Solution of the Generator Equation

We are faced with the situation that while the standard procedures for solving the differential equation and manipulating the resulting special functions gives a perfectly good formal solution, it is not of much practical use and something more creative is called for.

A key element in the alternative proposed here, is the generating function for Hermite polynomials, given by Abramowitz and Stegun (1965) as

$$e^{2rs-s^2} = \sum \frac{1}{n!} s^n H_n(r) . \tag{7.64}$$

Consider first the decelerating flow case. Substituting equation (7.59) into (7.63) we may write P^- explicitly as a Hermite expansion

$$\begin{aligned} P_T^-(z \mid z',0) &= e^{-z^2} \sum_n p_n H_n(z), \\ p_n &= \left(2^n n! \sqrt{\pi}\right)^{-1} e^{-nT} H_n(z'). \end{aligned} \tag{7.65}$$

Now consider a power series in an indeterminate variable a, using the same coefficients p_n:

$$\begin{aligned} \sum_n p_n(-a)^n &= \frac{1}{\sqrt{\pi}} \sum_n \frac{1}{n!}\left(-\frac{1}{2} a e^{-T}\right)^n H_n(z'), \\ &= \frac{1}{\sqrt{\pi}} e^{-a z' e^{-T} - \frac{1}{4} a^2 e^{-2T}} . \end{aligned} \tag{7.66}$$

Here, the second step follows by application of equation (7.64). Now, we rewrite the left hand side of equation (7.66) as

$$\begin{aligned} \sum_n p_n(-a)^n &= \sum_n \sum_m p_n(-a)^n \delta_{nm}, \\ &= \sum_n \sum_m p_n(-a)^n \frac{1}{\sqrt{\pi}\, m! 2^m} \int_{-\infty}^{\infty} e^{-z^2} H_n(z) H_m(z)\, dz. \end{aligned} \tag{7.67}$$

where the second step follows from the orthogonality relation of the $h_n(z)$. Next, we apply first equation (7.65) to the n-summation and then equation (7.64) to the m-summation:

$$\sum_n p_n(-a)^n = \frac{1}{\sqrt{\pi}}\int_{-\infty}^{\infty} dz\, P_T^-(z\,|\,z',0)\sum_m \frac{1}{m!}\left(-\frac{1}{2}a\right)^m H_m(z),$$
$$= \frac{1}{\sqrt{\pi}}\int_{-\infty}^{\infty} dz\, P_T^-(z\,|\,z',0)\, e^{-a\,z-\frac{1}{4}a^2}. \tag{7.68}$$

Finally, setting the right hand sides of equations (7.66) and (7.68) equal, we obtain

$$\int_{-\infty}^{\infty} dz\, e^{-az} P_T^-(z\,|\,z',0) = e^{-az'e^{-T}+\frac{1}{4}a^2(1-e^{-2T})}. \tag{7.69}$$

This is an integral equation that is very reminiscent of equation (7.38), and by comparison with how that equation arose from the Dynkin equation for constant flow we can conjecture that there should be a solution to the generator equation for decelerating flow of the form

$$f^-(z,T) = e^{-aze^{-T}+\frac{1}{4}a^2(1-e^{-2T})}. \tag{7.70}$$

Direct substitution of equation (7.70) into (7.55) confirms that is indeed a solution of the generator equation. Unlike the case of the Hermite polynomial solution, this solution does not allow the use of a closure relation to solve the integral equation; but instead, we model its solution on the one used for constant flow and that relied on completion of squares, by writing

$$P_T^-(z\,|\,z',0) = A\,e^{-(Bz+C)^2} \tag{7.71}$$

where A,B and C are terms independent of z. Substituting this expression in (7.69), completing squares and performing the integral by use of equation (7.41) leads to exponential expressions on both sides of the equation in which coefficients of powers of the arbitrary variable a can be set equal to fix the values of A,B and C. This results in a new expression for P^- :

$$P_T^-(z\,|\,z',0) = \frac{1}{\sqrt{\pi}\sqrt{1-e^{-2T}}}\exp\left(\frac{-(z-z'e^{-T})^2}{1-e^{-2T}}\right). \tag{7.72}$$

The same procedure applied to the case of accelerating flow leads to a somewhat different solution of the generator equation, but a rather similar expression for the probability density:

$$f^{+}(z,T) = e^{-[T+z^2+a z e^{-T}+\frac{1}{4}a^2(1-e^{-2T})]}, \tag{7.73}$$

and

$$P_T^{+}(z \mid z',0) = \frac{1}{\sqrt{\pi}\sqrt{e^{2T}-1}} \exp\left(\frac{-(z-z'e^{T})^2}{e^{2T}-1} \right). \tag{7.74}$$

In deriving equivalent expressions for the two probability densities, we have as a by-product also found a formula for performing the infinite sum over Hermite products contained in (7.63). Explicitly, the sum formula is

$$\sum_{n=0}^{\infty} \frac{e^{-nT}}{n!2^n} H_n(z) H_n(z') = \frac{e^{T+z'^2}}{\sqrt{e^{2T}-1}} e^{\frac{-(z-z'e^{T})^2}{e^{2T}-1}} . \tag{7.75}$$

It is directly evident that this formula holds by comparing the first equation in (7.63) with (7.74); it is less obvious that (7.75) also guarantees the equivalence of the second equation in (7.63) and (7.72), but that this true is a consequence of the following relationship between the probability densities for accelerating and decelerating flows:

$$P_T^{-}(z \mid z',0) = e^{T-z^2+z'^2} P_T^{+}(z \mid z',0). \tag{7.76}$$

This relationship is in turn self-evident from equations (7.63) but to prove that it also holds for the exponential expressions in equations (7.72) and (7.74) requires some algebraic manipulation.

It is instructive to see how the limit in equation (7.75) is approached, i.e. when the series is truncated at consecutively larger index values N, how well the partial sums represent the limiting function on the right hand side of the equation. In Figure 7.5, we show some plots of this comparison where the exponential expression is drawn as a continuous line and the partial sum as dashed line. Both functions are plotted as functions of z, for the fixed values $z' = 1$ and $T = 2$. From equation (7.75) it is clear that the limiting form is just a Gaussian peak if T is fixed, and the plots show that the partial sum represents the peak quite accurately for small values of $|z|$, but deviates drastically outside the range. As more terms are added to the sum the usable range is extended, but clearly large values of N are needed to give an acceptable representation over the full extent of the peak. The situation becomes worse if T is increased, because as is clear from the exponential form the position of the peak value is at $z = z'e^T$ which rapidly shifts to larger z values as T increases.

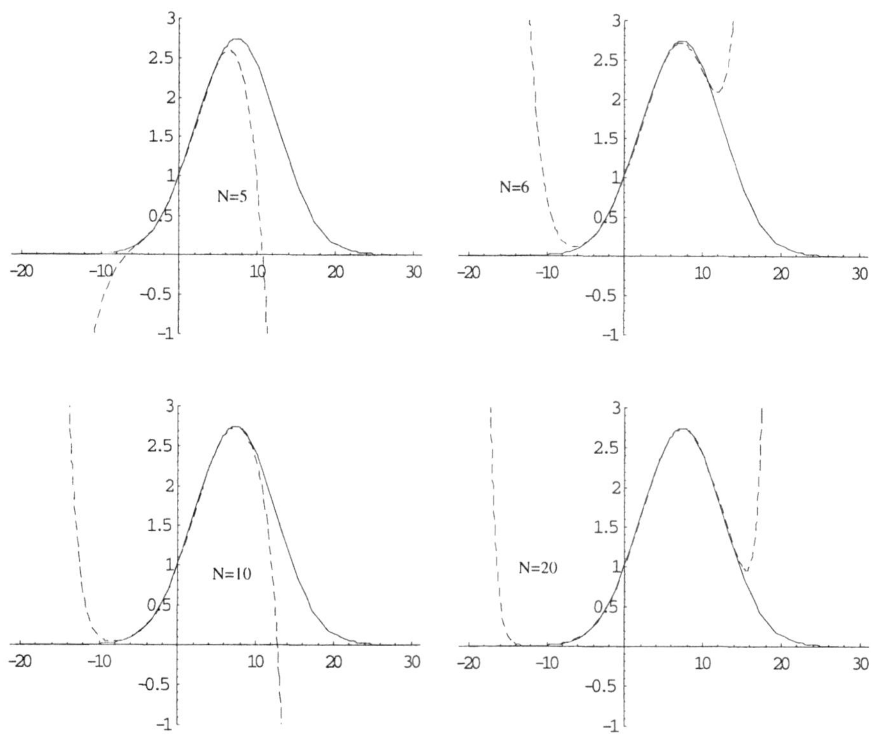

Figure 7.5 Comparisons of exponential expression and the partial sums.

This behavior demonstrates why the formally correct solution first found is not useful in practice, and may also boost the confidence of the skeptical reader about the correctness of the alternative solution in equations (7.72) and (7.74)!

For some purposes it is useful to collect the two probability expressions together by including the sign of the velocity gradient in the definition of the scaled time parameter T . In other words, also allowing for an arbitrary time origin, we may write

$$T = p(t - t') \tag{7.77}$$

in which case

$$P_T^{\pm}(z \mid z', 0) = \frac{1}{\sqrt{\pi}\sqrt{\pm(e^{2T} - 1)}} \exp\left(\frac{-(z - z'e^T)^2}{\pm(e^{2T} - 1)}\right). \tag{7.78}$$

As additional confirmation of this expression, one may check by straightforward algebra that as required by physical considerations, it is mathematically reduced to the appropriate limits:

- in the limit $T \rightarrow 0$, the probability density reduces to $\delta(z-z')$;
- in the deterministic limit, substituting in equations (7.54) and letting $\gamma \rightarrow 0$, it reduces to $\delta(x-X(t))$ where $X(t)$ is given by equation (7.23) with the source point $(\xi, 0)$ appropriately replaced by $(x', 0)$; and
- in the constant flow velocity limit, once more reverting from z and T to x and t, and then taking the limit $p \rightarrow 0$, equation (7.42) is retrieved.

7.7 Evolution of a Gaussian Concentration Profile

Before the time dependent concentration can be calculated, we still need to transform the target-point probability density $P(z\,|z0,0)$ found above, to the source-point density $P(z0\,|z,T)$. The procedure is the same as was discussed for a constant flow velocity; the only difference is that now the variable y (given in that case by equation (7.46)) characterizing the joint source-target probability density, has the more complicated form as implied by equation (7.78) :

$$y = \frac{z - z' e^{T}}{\sqrt{\pm(e^{2T} - 1)}} . \tag{7.79}$$

From the form of equation (7.79) it is evident that when applying equations (7.45) the expression for $P_T(z|z',0)$ in (7.78) will be recovered but that $P_0(z'|z,0)$ differs from it by a factor e^T , allowing for the necessary exchange of integration limits:

$$P_0^{\pm}(z' \mid z,T) = \frac{e^{T}}{\sqrt{\pi}\sqrt{\pm(e^{2T}-1)}} \exp\left(\frac{-(z - z' e^{T})^2}{\pm(e^{2T}-1)} \right). \tag{7.80}$$

The importance of this factor, and in fact the need to distinguish between the two closely related probabilities that in the constant flow case reduced to the same expression, can be well illustrated by considering integrals over the infinite z- and z'-domains. Using equation (7.79) as an integration substitution it is easy to see that

$$\int_{-\infty}^{\infty} dz\, P_T(z \mid z',0) = 1; \quad \int_{-\infty}^{\infty} dz'\, P_T(z \mid z',0) = e^{-T} . \tag{7.81}$$

Formulated verbally, the first of these equations expresses the fact that given the presence of a fluid element at the source point $(z',0)$ fluid mass conservation ensures that it must be found at some target point z at the later time T; so summing the probabilities over all possible target points gives 1. On the other hand, if we select any target point z at random, there is no guarantee that there exists a source point from which a fluid element will proceed to z – it is logically conceivable that some points are unreachable, and in such a case summing probabilities over all source points will give 0. The second equation in equation (7.81) shows that indeed for accelerating flow this probability sum is not 1. But if we formulate the statement in terms of the different conditional probability of equation (7.80), it reads: Given that a fluid element is found at the target point *(z,T)*, it must have come from somewhere at *T=0*; so summing this probability over all possible source points must once again give 1. Obviously, the extra factor e^T in equation (7.80) is just what is needed to ensure that that happens.

Finally, we are in a position to calculate the evolution of an input concentration profile in an accelerating flow from equation (7.26) by substituting a Gaussian input peak from equation (7.20), the flow velocity from equation (7.53) and the probability density from (7.80) re-expressed in terms of x and t. The resulting integrand is complicated, but a moment's consideration shows that the terms in the exponent contributed by the concentration and the probability are both quadratic expressions in the integration variable x', so we can write the integral as

$$
\begin{aligned}
c(x,t) &= N\int_{-\infty}^{\infty} dx' u(x')\exp-[A x'^2 + B x' + C], \\
N &= \frac{\mu / \gamma}{s\sqrt{2\pi^2}\, u(x)\sqrt{\pm(e^{2T}-1)}}.
\end{aligned}
\tag{7.82}
$$

where A,B and C are complicated expressions but independent of x' . Squares in the exponent are completed by transforming to an integration variable $Y = x' + B/2A$, in terms of which

$$
c(x,t) = N\exp[\frac{B^2}{4A} - C]\left\{\left(v_0 - p\frac{B}{2A}\right)\int_{-\infty}^{\infty} dY\,\exp[-AY^2] + p\int_{-\infty}^{\infty} dY\; Y\exp[-AY^2]\right\}
\tag{7.83}
$$

By inspection of equations (7.20) and (7.80) it is seen that as the coefficients of the quadratic term in each is negative, A must be positive and therefore the second integral in the curly brackets vanishes while the first reduces to $\sqrt{\pi / A}$. All that remains at this point is to construct the detailed expressions for A,B and C and simplify the combinations of these terms that occur in (7.83), a task best accomplished by the use of symbolic algebra software. As in the deterministic case, we take $x_0 = \xi$ to simplify the algebra. The result is as follows, expressed in terms of the auxiliary variable $\phi = \gamma^2 / \mu^2$:

$$
c(x,t) = M(x,T)\,\mathcal{G}\left(x - X(T),\; s^2 e^{2T} \pm \tfrac{1}{2}\phi(e^{2T}-1)\right)
\tag{7.84}
$$

where $\mathcal{G}$ is the normalised Gaussian, *X(T)* is as given by equation (7.23) and *M* is a modulation factor:

$$\mathcal{G}(x,\sigma)=\frac{1}{\sigma\sqrt{2\pi}}\exp\left[\frac{-x^2}{2\sigma^2}\right], \tag{7.85}$$

$$M(x,T)=\frac{2s^2u(x)\pm\phi(e^{2T}-1)e^Tu(\xi)}{u(x)\left[2s^2e^{2T}\pm\phi(e^{2T}-1)\right]}. \tag{7.86}$$

Clearly, the propagation of a concentration plume is much more complex in an accelerating flow than in the constant flow velocity previously discussed. Even so, one can relate several of its features to those of constant velocity dispersion. Most striking is that as before there is a Gaussian peak and it also remains centered on the deterministic trajectory $X(t)$, as was the case for both equation (7.24) (accelerating flow, no dispersion) and equation (7.47) (constant velocity, with dispersion). The main difference is that here the variance of the Gaussian is given by

$$\sigma^2(T)=s^2e^{2T}\pm\tfrac{1}{2}\phi(e^{2T}-1). \tag{7.87}$$

In the deterministic limit, $\gamma\to 0$, so $\phi\to 0$ and only the first term remains in agreement with equation (7.25). This means that the first term represents kinematical dispersion while the second one represents the intrinsic stochastic dispersion. Indeed, in the constant velocity limit $p\to 0$ it is seen that the intrinsic dispersion term reduces to the diffusive dispersion expression (7.49) as it should.

We also need to consider the modulation factor by which the Gaussian is multiplied. Bearing in mind that with the choice made for x_0 we have $u(\xi)=v_0$, it follows from equation (7.23) that

$$u(\xi)e^T=u(x)-p(x-X(T)) \tag{7.88}$$

and hence together with equation (7.87) we find

$$M(x,T)=1-\frac{\pm\phi(e^{2T}-1)\,p(x-X(T))}{2\sigma^2(T)u(x)}. \tag{7.89}$$

Consider the behavior of M at a fixed time value. For x at the (moving) position of the peak we see from equation (7.89) that it has the value 1. Provided that there is no stagnation point (i.e., $u(x)=0$) in the x-interval under consideration, a case which is not generally of physical interest and was excluded from consideration already when discussing deterministic flow, both the numerator and denominator of the second term vary linearly as we move away from the peak and so the value of M will only change slowly with x – in particular compared to the Gaussian which falls away exponentially to either side of the peak. Also, note that deviations of M from 1 are proportional to the velocity gradient, so they fall away for a constant flow and a strictly Gaussian peak shape is recovered in that case. Generally it is reasonable to describe the concentration peak as still essentially Gaussian, but with a moderate time varying modulation. The explicit time variation in equation (7.89) will also largely cancel between the numerator and denominator. However, there is also an implicit time variation caused by the velocity factor that contributes mainly at the x position of the peak, and this increases with time. That will tend to extinguish the modulation with time for an accelerating flow, while for decelerating flow a slow increase will result.

The most striking qualitative feature of the modulation is that it is asymmetric, because the second term in equation (7.89) undergoes a sign change when x is taken through the peak value at $X(T)$. So flow acceleration causes the dispersion peak to become skewed, because values on one side of the peak is enhanced by the modulation and reduced on the other side; the direction of skewing will be backwards for acceleration and forwards for deceleration. Once more no drastic effect is expected because the numerator and denominator changes in a coordinated way.

In fact, on actual numeric plots the skewing is hard to recognize visually. Nevertheless the phenomenon is significant in principle, because it is an example of the fact that dispersion in the presence of a velocity gradient is different from merely superimposing stochastic variation on the deterministic evolution. We have seen in equations (7.24) and (7.47) that taken separately, the effects of flow acceleration and dispersion produce perfectly symmetric Gaussian concentration peaks; and yet we find here that together they give rise to a skewed, quasi-Gaussian peak. This is another demonstration of the essential difference between SDE's and the addition of random effects to a deterministic differential equation, that was discussed in some detail in Chapter 5 in connection with the population growth model.

The same point is made in a more dramatic fashion by considering in more detail the time behavior of the plume variance given in equation (7.87).

Separately, stochastic dispersion produces a variance that increases linearly with time (equation (7.48)) and flow acceleration gives an exponential increase (equation (7.25)). Together, this same exponential increase is present as the first term in equation (7.87), but in addition the stochastic term also acquires its own exponential time dependence. Once more, simple superposition fails and we might describe the result as an interaction between dispersion and the flow acceleration that is described by the full SDE treatment.

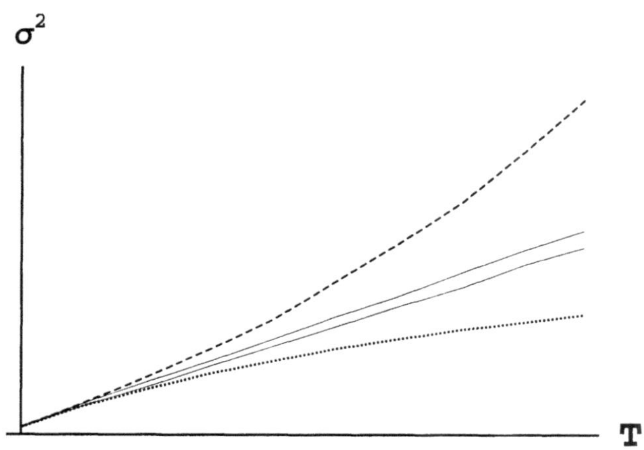

Figure 7.6 Plume variance in accelerating and decelerating flows.

The extent of the interaction may be judged by inspecting the plot in figure 7.6. It shows the calculated plume variance according to equation (7.87) in accelerating flow as a dashed line and in a decelerating flow as a dotted line. For comparison, the two solid lines show the result that would be obtained by superimposing a linear (diffusive) dispersion on two kinematical dispersion rates, for the same acceleration and deceleration rates respectively. The relatively small separation of the solid lines shows that for the parameters chosen for the plot, kinematical dispersion is quite small compared to the diffusive effect that would hold in a constant flow regime over the same time interval. Nevertheless, the interaction of this small kinematical effect and the dispersion produces a much larger final effect, enhancing dispersion for acceleration and suppressing it for deceleration.

The non-linear time dependence is also very significant from the point of view of traditional models of solute transport. In fact, the usefulness of the concept of dispersivity is called into question. One might formally extract a time dependent dispersivity expression by taking the time derivative of equation (7.87), but it does not reduce to a pure materials property that describes the porous medium any more, as it did for the constant flow expression discussed after equation (7.52). Instead, not only time, but also other variables such as the initial plume extension represented by s and the velocity gradient p that may at least partially be determined by the flow geometry, occur. It is more consistent to concede that when the flow velocity varies the Fickian model of dispersion is not adequate any more, and has to be replaced by a more complex model to describe the evolution of a contaminant plume, and that is exactly what equation (7.87) does for the case of linear acceleration.

On the other hand, if we do maintain the definition of dispersivity for the sake of comparison with the traditional description, the fact that it becomes time dependent opens the way to an explanation of the "scale dependence" observed in practical dispersivity measurements, as was discussed in Chapter 1. When the spatial scale of an experiment is increased, the time interval over which dispersion is observed also increases, and so scale- and time dependence are just different ways to describe the same phenomenon.

Having said that, the results derived here are still a long way from giving a detailed explanation of the observed scale effect. In fact, the mathematical treatment in this section would apply to a semi-infinite range of linearly increasing flow velocity, limited only on one side by the presence of a stagnation point. This is not a physically realistic situation. That is clear, for example, from the presence of exponential growth for a positive acceleration, which was found in both the deterministic and stochastic solutions. Obviously such exponential growth in any physical quantity is physically sustainable only over a limited spatial or time range. Indeed, the physical examples discussed at the beginning of this section show a linear velocity profile only over comparatively short spatial intervals. And in the physically realistic situation where there is a stable average flow velocity over larger scales, one would expect that intervals of velocity growth would alternate with intervals where the velocity decreases, i.e. velocity fluctuation rather than a sustained velocity growth is physically relevant.

If the results derived in this section are to be applied to such problems, it would thus have to be done by joining together solutions over discrete intervals in a piecewise linear model of the flow velocity. That leads to the necessity of solving boundary problems at the points where intervals join, and that turns out to be a much more demanding problem mathematically than the initial value problem solved above.

The observation that one can make from Figure 7.6 that dispersion tends to be enhanced by acceleration and suppressed by deceleration, may lead one to speculate that a key issue in such a treatment will be whether these effects cancel over the extent of a fluctuation. If so, the ADE-type approach where it is assumed that dispersion may be described in terms of an average flow velocity should be in order; but if not, additional effects from fluctuations will change the dispersion behavior and might give rise to the observed scale effects.

In fact, it is easy to see from conservation of flux that kinematical dispersion is reversible and must cancel over the length of a fluctuation. In a simplified situation the form of equation (7.87) readily demonstrates this. A Gaussian peak that starts from an initial variance s^2 and propagates for a time τ in a flow with acceleration coefficient μ^2 in the absence of stochastic effects attains a variance value of $s^2 e^{2\mu^2\tau}$ according to (7.87). If at that moment the acceleration is reversed, the variance after a further time interval τ will by the same argument be given by $\left(s^2 e^{2\mu^2\tau}\right)e^{-2\mu^2\tau} = s^2$, i.e. the stretching of the peak during the first phase is exactly reversed in the second phase. On the other hand, the functional form of the intrinsic dispersion term does not allow the same manipulation, showing explicitly that that part is not reversible. Of course, a plume that consecutively penetrates an accelerating and a decelerating region is not quite the same as the hypothetical reversal of the acceleration at a given time, but if the intrinsic dispersion is not even reversible in the simpler situation it is plausible that it will also not cancel over the extent of a velocity fluctuation.

To summarize in conclusion, in addition to the detailed results on dispersion in constant and linearly accelerating flows, the study in this chapter also allows some observations of broader significance. Firstly, it demonstrates how an SDE-based model introduces interaction between the stochastic and deterministic aspect of a physical system. Secondly, the result in the contaminant dispersal problem is to modify the time dependence of dispersion, compared to the case where the interaction is neglected. Lastly, on a qualitative level at least, this modification has a bearing on explaining the observed scale dependent behavior of dispersivity. As conventional models do not explain this behavior, pursuit of the SDE approach is a promising avenue for further research.

Chapter 8

A Stochastic Computational Model for Solute Transport in Porous Media

8.1 Introduction

Computational models can often be used to investigate the phenomena it describes through experimentation with the model. In this chapter, we develop a model that describes the solute dispersion in a porous medium saturated with water considering velocity of the solute as a fundamental stochastic variable. When we consider the hydrodynamic dispersion of a solute in flowing water in a porous medium, there are two ways the solute gets distributed over the medium. The solute can mechanically disperse due to fingering effects of the granular medium and it can diffuse due to solute concentration differences. In deriving the advection-dispersion equation for solute transport, the dispersive transport is modeled by using a Fickian assumption which gives rise to the hydrodynamic dispersion coefficient (Fetter, 1999). We have seen how the perturbation term of the velocity, gives rise to the dispersive flux in Chapter 1 (section 1.3.2) and then the flux term is assumed to be related to the concentration gradient on plausibility arguments. The hydrodynamic dispersion coefficient has been found to be dependent on the scale of the experiment. The hydrodynamic dispersion contributes to making the velocity of solute particles a random variable by changing direction and magnitude in an unpredictable manner. In this chapter, we develop a model that addresses this fundamental nature of the dispersion phenomenon in porous media.

The basic assumption on which this model is based is that the velocity of solute particles is fundamentally a stochastic variable, with irregular but continuous realizations. Given the observations of Rashidi et al. (1996), this is a reasonable hypothesis. If we hypothese that the velocity constitutes of component representing mean and/or a typical value, and a fluctuating component which depends on the characteristics of the irregular geometry of the porous medium, we can express velocity similar to equation (6.1):

$$\underline{v}(\underline{x},t) = \bar{v}(\underline{x},t) + \underline{w}(\underline{x},t). \tag{6.1}$$

$\bar{v}(\underline{x}, t)$ can be expressed as a Darcian description or as a "typical" average value of velocity for the region considered. However, we need to understand the interplay between $\bar{v}(\underline{x},t)$ and $\underline{w}(\underline{x},t)$, and the spatial region can change the value of $\bar{v}(\underline{x},t)$, and in extreme heterogeneous case, we can use $\underline{w}(\underline{x},t)$ to model the velocity recognizing that $\bar{v}(\underline{x},t)$ may not meaningfully exist. Once we have recognized this, the variables such as solute flux associated with solute transport must be treated as stochastic variables, and we need to derive the mass conservation for the solute transport problem based on theories in stochastic calculus.

8.2 Development of a Stochastic Model

We proceed to develop the model using the simplest setting possible, and for this purpose we use 1-D domain [0,a] and we keep solute concentration at x=0 at a constant value. Let us consider a 1-dimensional problem of a solute dispersion in a saturated porous medium. Consider concentration $C(x,t)$ as a stochastic variable with , for example, g/m^3 as units, $V(x,t)$ is the velocity (m/h), φ is the porosity of the material and $J(x,t)$ is the contaminant flux at x in $g/m^2.h$. As C, V, and J are stochastic functions of space and time having irregular (sometimes highly irregular) and continuous realizations, it is important to consider higher order terms to the Taylor series expansion when formulating the mass conservation model for the solute. Consider an infinitesimal cylindrical object having a cross sectional area, A (Figure 8.1).

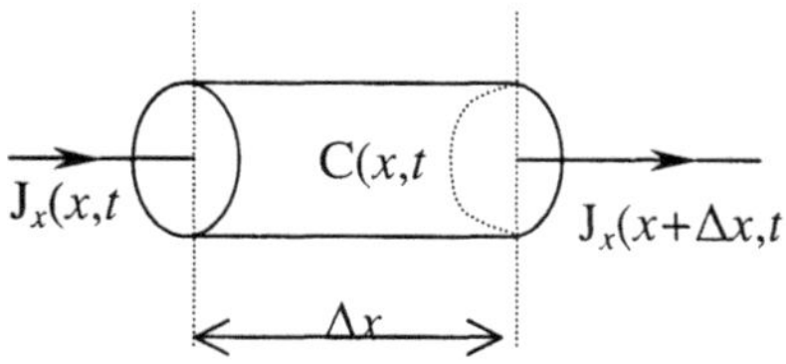

Figure 8.1 An infinitesimal cylindrical object within the porous medium having the solute concentration of C(x,t).

Writing the mass balance for the change in solute during a small time increment, Δt,

$$\Delta C(x,t)\ \varphi \mathrm{A}\ \Delta x = \left(J_x(x,t) - J_x(x+\Delta x,t)\right)\ \varphi \mathrm{A}\ \Delta t,$$

$$\left(\frac{\Delta C}{\Delta t}\right)_{x,t} = \frac{\left(J_x(x,t) - J_x(x+\Delta x,t)\right)}{\Delta x}. \tag{8.1}$$

For convenience, let us indicate $J_x(x,t)$ as J_x and $J_x(x+\Delta x,t)$ as $J_{x+\Delta x}$.
From the Taylor series expansion,

$$J_{x+\Delta x} - J_x = \frac{1}{1!}\frac{\partial J_x}{\partial x}\ \Delta x + \frac{1}{2!}\frac{\partial^2 J_x}{\partial x^2}\ (\Delta x)^2 + \frac{1}{3!}\frac{\partial^3 J_x}{\partial x^3}\ (\Delta x)^3 + R(\varepsilon)$$

where R(ε) is the remainder of the series.

Assuming that the higher order derivatives greater than 3 of the flux are negligible, equation (8.1) can be written as

$$\frac{\partial C}{\partial t} = -\frac{\partial J_x}{\partial x} - \frac{1}{2}\frac{\partial^2 J_x}{\partial x^2} dx + R_c(x,t), \tag{8.2}$$

where,

$$R_c(x,t) = -\frac{1}{6}\frac{\partial^3 J_x}{\partial x^3}(dx)^2 - \frac{1}{24}\frac{\partial^4 J_x}{\partial x^4}(dx)^3 -$$

As the flux changes with the velocity, J can also be considered as changing rapidly within the porous medium creating dispersion of solutes. Even a slight variation of velocity would change the dispersion significantly over time; therefore, it is important to keep the non-linearity of the phenomenon intact if possible. We assume that the second order term captures non-linear behavior sufficiently. However in extreme heterogeneous situations, more higher order terms are needed to develop the computational model.

Substituting $dx = h_x$,

$$\frac{\partial C}{\partial t} = -\frac{\partial J_x}{\partial x} - \frac{h_x}{2}\frac{\partial^2 J_x}{\partial x^2} + R_c(x,t). \tag{8.3}$$

Equation (8.3) describes the mass conservation of the contaminant within the cylindrical volume ($A\Delta x$). Here we circumvent the discussion of nature of the variables. Obviously, $C(x,t)$ is an average over the cylindrical volume, but by making Δx smaller and smaller, we can think of $C(x,t)$ as a pore scale concentration representing a small region in space, but, if Δx is in the same order of magnitude as a typical grain size of the porous medium under consideration, $C(x,t)$ loses its meaning, so use of a realistic Δx for the medium considered is important.

$$dC = -\left(\frac{\partial J_x}{\partial x} + \frac{h_x}{2}\frac{\partial^2 J_x}{\partial x^2}\right) dt + R_c(x,t)\ dt.$$

Compared to the first term on the right hand side, let us assume that $R_c(x,t)dt \approx 0$. This assumption has to be tested in any given situation.

$$dC = -\left(\frac{\partial J_x}{\partial x} + \frac{h_x}{2}\frac{\partial^2 J_x}{\partial x^2}\right) dt. \tag{8.4}$$

Let us express the $J(x,t)$ term in terms of the velocity in the x direction and the concentration of the contaminant:

$$J(x,t) = V(x,t)\ C(x,t)\,. \tag{8.5}$$

Now the velocity can be expressed as a stochastic quantity which is affected by the nature of the porous medium. The effects of the porous medium can be included within the noise term of the stochastic variable. We model the velocity in terms of the mean velocity and the Gaussian white noise:

$$V(x,t) = \bar{V}(x,t) + \xi(x,t)\,. \tag{8.6a}$$

For the purpose of simplicity, we assume that $\bar{V}(\underline{x},\ t)$ exists and we can express $\bar{V}(\underline{x},t)$ using the Darcy's Law:

$$\bar{V}(x,t) = -\frac{K(x)}{\varphi(x)}\frac{\partial p}{\partial x} \qquad \text{(Darcy's Law)} \tag{8.6b}$$

where

$K(x) =$ a typical value of the hydraulic conductivity in the region,
$\varphi(x) =$ the porosity of the material, and
$p =$ pressure head .

$\xi(x,t)$ is white noise correlated in space and δ-correlated in time such that

$$\mathrm{E}[\xi(x,t)] = 0 \tag{8.7}$$

$$\mathrm{E}[\xi(x_1,t_2)\ \ \xi(x_2,t_2)] = q(x_1,x_2)\,\delta(t_1 - t_2) \tag{8.8}$$

$q(x_1,x_2)$ is the velocity covariance function in space and $\delta(t_1 - t_2)$ is the Dirac's delta function.

We can express equation (8.8), substituting $\Delta t = t_2 - t_1$,

$$\mathrm{E}[\xi(x_1,t_1)\ \ \xi(x_2,t_1 + \Delta t)] = q(x_1,x_2)\delta(\Delta t)\,. \tag{8.8a}$$

The Dirac's delta function $\delta(\tau)$ is defined so that it is zero everywhere except at $\tau = 0$ when it is infinite in such a way that

$\int_{-\infty}^{\infty} \delta(\tau)\, d\tau = 1$.

In equation (8.8a), we separate the velocity into two separate spatial and temporal components, and therefore, we say that $\xi(x,t)$ is correlated in space and δ- correlated in time. In practice, this means that we can include a "description" of porous medium through $q(x_1, x_2)$ and keep essentially the irregular behavior with respect to time. However, one could argue for the case where the velocity noise term ($\xi(x,t)$) is correlated both in space and time through covariance functions. In this case the difficulties in solving the subsequent mathematical and computational problems within the context of applications can be insurmountable. In addition, the correlation in space and δ – correlated in time of $\xi(x,t)$ can be justified based on plausible grounds. It allows us to model the randomness induced by the irregular geometry of the medium directly and we will be able to focus on the effects of heterogeneity of the medium.

Proceeding with the mathematical derivation by substituting equation (8.6a) into equation (8.5),

$$J(x,t) = \left(\bar{V}(x,t) + \xi(x,t)\right)\ C(x,t), \tag{8.9}$$

$$J(x,t) = \bar{V}(x,t)C(x,t) + C(x,t)\xi(x,t). \tag{8.10}$$

Substituting equation (8.10) into equation (8.5),

$$\begin{aligned} dC &= -\frac{\partial}{\partial x}\left[\bar{V}(x,t)C(x,t) + C(x,t)\xi(x,t)\right] - \frac{h_x}{2}\frac{\partial^2}{\partial x^2}\left[\bar{V}(x,t)C(x,t) + C(x,t)\xi(x,t)\right], \\ &= -\left(\frac{\partial}{\partial x} + \frac{h_x}{2}\frac{\partial^2}{\partial x^2}\right) * \left[\bar{V}(x,t)C(x,t) + C(x,t)\xi(x,t)\right]. \end{aligned} \tag{8.11}$$

Let us define the operator in space,

$$S = -\left(\frac{h_x}{2}\frac{\partial^2}{\partial x^2} + \frac{\partial}{\partial x}\right) \text{ for a given } h_x.$$

Then:

$$dC = S\left(\bar{V}(x,t)C(x,t) + C(x,t)\xi(x,t)\right)dt\,,$$

$$dC = S\left(\bar{V}(x,t)\ C(x,t)\right)\ dt + S\left(C(x,t)\ \left(\xi(x,t)\ dt\right)\right). \tag{8.12}$$

Equation (8.12) has the form of a stochastic differential equation and both terms on the right hand side can be integrated as Ito integrals to obtain concentration. We introduce $d\beta(t) = \xi(x,t)dt$ where $\beta(t)$ is a Wiener process in Hilbert space for a given x.

Therefore equation (8.12) can be written as:

$$dC = S\left(\bar{V}(x,t)\ C(x,t)\right)\ dt + S\left(C(x,t)\ d\beta(t)\right). \tag{8.13}$$

This means in the integral notation,

$$C(x,t) = \int_o^t S\left(\bar{V}(x,t)\ C(x,t)\right)\ dt + \int_o^t S\left(C(x,t)\ d\beta(t)\right) \tag{8.14}$$

where S is the differential operator given above.

Unny (1989) showed that $d\beta(t)$ can be approximated by:

$$d\beta_m(t) = \sum_{j=1}^{m} f_j\sqrt{\lambda_j}\ \ db_j(t)\,, \tag{8.15}$$

where m is the number of terms used, $db_j(t)$ is the increments of standard Wiener processes, f_j and λ_j are eigen functions and eigen values of the covariance function of the velocity, respectively.

8.3 Covariance Kernel for Velocity

Ghanem and Spanos (1991) describe the mathematical details of expressing the noise term of a stochastic variable (e.g. velocity in this case) as a Karhunnen-Loeve expansion. The central to this expansion is the choice of the covariance function (Covariance Kernel) which models the spatial correlation of the 'noise' term ($\xi(x,t)$ in equation (8.6a)). We assume an exponential covariance kernel in this work to illustrate the model development. The exponential covariance kernel is frequently used in modeling the correlation of geographical data.

The exponential covariance kernel can be given as:

$$q(x_1, x_2) = \sigma^2 e^{\frac{-y}{b}} \tag{8.16}$$

where $y = |x_1 - x_2|$, b is the correlation length and σ^2 is the variance (Ghanem and Spanos, 1991). x_1 and x_2 are any two points within the range [0,a]. The eigen functions (f_n) and eigen values (λ_n) of $q(x_1, x_2)$ are obtained as the solution to the following integral equation:

$$\int_0^a q(x_1, x_2)\ f_n(x_2)\ dx_2 = \lambda_n f_n\ (x_1)\,. \tag{8.17}$$

The solutions to equation (8.17) assuming σ^2 is a constant over [0,a] are given by:

$$\lambda_n = \frac{2\theta\sigma^2}{\omega_n^2 + \theta^2}, \tag{8.18}$$

where $\theta = 1/b$ and ω_n's are the roots of the following equation:

$$\tan \omega_n a = \frac{2\omega_n \theta}{\omega_n^2 - \theta^2}. \tag{8.19}$$

The orthonormal basis functions of the Hibert space associated with the exponential kernel are the eigenfunctions given by the integral equation (8.17). Equation (8.17) can be solved to yield the following function as the n th basis function:

$$f_n(x) = \frac{1}{\sqrt{N}}\left(\sin \omega_n x + \frac{\omega_n}{\theta} \cos \omega_n x \right), \tag{8.20}$$

where

$$N = \frac{1}{2} a \left(1 + \frac{\omega^2}{\theta^2}\right) - \frac{1}{4\omega}\left(1 + \frac{\omega^2}{\theta^2}\right) \sin 2\omega\ a - \frac{1}{2\theta}(\cos 2\omega\ a - 1). \tag{8.20a}$$

8.4 Computational Solution

8.4.1 Numerical Scheme

The differential operator S in equation (8.12) can be expressed as a difference operator using a backward difference scheme. By dividing the interval from 0 to a on x axis into ($\underline{N}$-1) equidistant and small intervals of Δx, and the interval from 0 to t on the time axis into ($\underline{K}$-1) equidistant and small intervals of Δt, we can write the derivatives for any variable U at (k,n) point on the space-time grid (Figure 8.2).

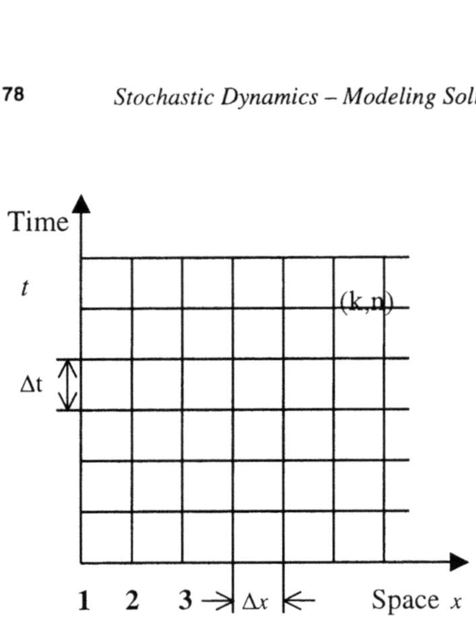

Figure 8.2 Space-time grid used in the computational scheme with respect to x.

The first derivative of a variable U can be written as

$$\left(\frac{\partial U}{\partial x}\right)_k^n = \frac{U_k^n - U_{k-1}^n}{\Delta x} \tag{8.21}$$

where U_k^n indicates the value of U at the grid point, (n,k). The second derivative can be written as

$$\left(\frac{\partial^2 U}{\partial x^2}\right)_k^n = \frac{U_k^n - 2U_{k-1}^n + U_{k-2}^n}{\Delta x^2}. \tag{8.22}$$

The operator S can be written as,

$$SU = -\left(\frac{\partial U}{\partial x} + \frac{h_x}{2}\frac{\partial^2 U}{\partial x^2}\right).$$

In the difference form

$$(SU)_k^n = -\left(\left(\frac{\partial U}{\partial x}\right)_k^n + \frac{h_x}{2}\left(\frac{\partial^2 U}{\partial x^2}\right)_k^n\right).$$

Substituting from equation (8.21) and equation (8.22) and taking $h_x = \Delta x$,

$$(SU)_k^n = -\left(\frac{1}{2\Delta x}\right) * \left[3U_k^n - 4U_{k-1}^n + U_{k-2}^n\right]. \tag{8.23}$$

The first derivative of U with respect to time can be expressed using a forward difference scheme:

$$\frac{\partial U}{\partial t} = \frac{U_k^{n+1} - U_k^n}{\Delta t}. \tag{8.24}$$

Applying equation (8.23) and (8.24) to equation (8.23) and for the case of the mean velocity (v) being constant,

$$\begin{aligned} C_k^{n+1} = C_k^n &- \left(\frac{\Delta t\, v}{2\Delta x}\right) * \left[3C_k^n - 4C_{k-1}^n + C_{k-2}^n\right] \\ &- \left(\frac{1}{2\Delta x}\right) * \left[3C_k^n \beta_k^n - 4C_{k-1}^n \beta_{k-1}^n + C_{k-2}^n \beta_{k-2}^n\right] \end{aligned} \tag{8.25}$$

The difference equation (8.25) gives the future value of a stochastic variable in terms of past values. In addition explicit solution procedure possesses the properties of Ito definition of integration with respect to time.

The numerical solution was implemented in Mathematica taking the numerical convergence and stability into account. The scheme has extensively been tested for 1-dimensional case, and it was found that 30 terms in equation (8.15) give a very high degree of numerical accuracy in the solution. The Mathematica program is flexible to incorporate different boundary conditions, and it was designed to test the behavior of the model with in different b and σ^2 regimes. The details of the implementation is not discussed here but are available from the authors.

8.4.2 The Behavior of the Model

As an example, we have solved equation (8.19) with a = 1.0m correlation length, b = 0.05m and obtained 11 roots: ω_1 = 2.85774; ω_2 = 5.72555; ω_3 = 8.6116; ω_4 = 11.2511; ω_5 = 14.4562; ω_6 = 17.4166; ω_7 = 23.4054; ω_8 = 26.4284; ω_9 = 29.4669; ω_{10} = 32.5187 and ω_{11} = 35.5871. In this particular case, 11 terms in equation (8.15) is sufficient for extremely good numerical accuracy of the solution.

With these roots we have constructed the basis functions using equation (8.20). With $\sigma^2 = 1.0$ we have calculated the eigen values λ_n to construct the increments of Wiener processes in the Hilbert spaces using equation (8.15). The standard Wiener process increments were generated for $\Delta t = 0.0001$ days for a total time of 1 day (see Kloeden and Platen (1991)). The value of 50.0 m/day was used for the hydraulic conductivity and piezometric head gradient of 0.020 m/m was used to obtain the mean velocity of 4.0 m/day for a porous medium having porosity of 0.25. A realization of the solution is given in Figure 8.3.

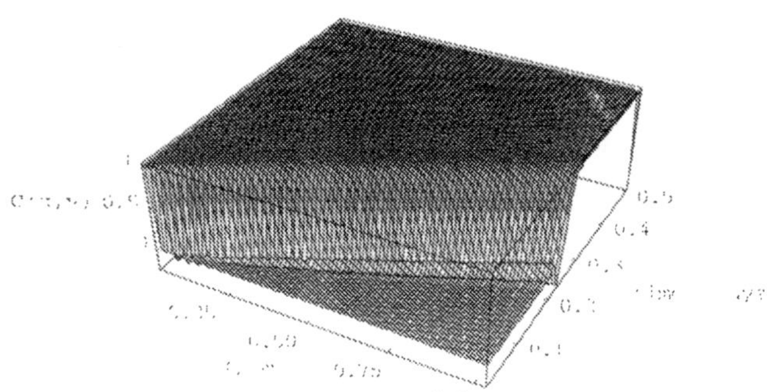

Figure 8.3 A realization of the computational solution.

The statistical nature of the computational solution changes as σ^2 and b change. This allows us to model hydrodynamic dispersion without the need for a scale dependent diffusion coefficient. In this example, we have a very high value for the mean velocity and as a result we can expect advection to dominate as seen in this example; therefore, high stochastic amplitude of 1.0 does not have a significant effect on the realization. Next we will use a simplistic setting to examine the behavior of the model to gain some insight as to how σ^2 and b influence hydrodynamic dispersion.

To illustrate the behavior of the model, we solved the 1-dimensional problem for the domain $0 \le x \le 1$ with a constant concentration of 1.0 applied at the boundary x = 0. Then we obtained the temporal development of the concentration profile at the mid point of the domain $x = 0.5$ for various combinations of σ^2 and b and we have kept all other parameters constant: the mean velocity was taken to be 0.5 m/day and we have used the same standard Wiener increments for all the experiments. Qualitative nature of our understanding about how the model behaves proved to be independent of the particular realization of the Wiener process.

8.5 Computational Investigation

We have investigated the Stochastic Solute Transport Model (SSTM) described in the previous section for simple settings of one-dimensional case to understand its behavior. The main parameters of the model are the correlation length, b and the variance, σ^2. As the statistical nature of the computational solution changes with b and σ^2, the main objective of this exercise is to identify effects of these parameters to the solution of the model.

The distributed concentration values were obtained by using the finite difference numerical solution taking the numerical convergence and stability into consideration. We first illustrate the behavior of the model by solving one-dimensional problem for the spatial domain of 1 m ($0 \le x \le 1$). We solved equation (8.19) to generate the roots for a given set of parameters. For an instance, for the correlation length, b = 0.1 m, we obtained 29 roots: ω_1 = 2.62768, ω_2 = 5.30732, ω_3 = 8.06714, ……, ω_{29} = 88.1904. Generally 30 terms are more than sufficient to produce converging numerical solutions. We generated the standard Wiener process increments in Hilbert space for the time intervals of 0.001 days for the total time of 3 days. Then eigenvalues λ_n were computed for the required σ^2. With these roots, ω and λ_n, we have then calculated the basis function in equation (8.20). Those values were used to generate $d\beta(t)$ in equation (8.15). The numerical scheme of SSTM was then used to calculate the concentration profile for spatial-temporal development for the mean velocity of 0.5 m/day. This value of mean velocity can be thought as large enough to have a sufficient plume development within a few days and small enough to depict the variability in concentration profiles.

We have used spatial grid length of 0.1 m for the numerical calculations. It can be shown by analyzing equation (8.14) mathematically, the grid size (h_x) less than 0.1 does not effect the solution significantly. Initial concentration value of 1.0 unit was considered at $x = 0$ and it was assumed as a continuous source for the entire time period of the solution. Exponentially distributed concentration values of $e^{-5k\Delta x}$, where $k = 1, 2, \ldots, 10$ and Δx is the grid size, were considered as the initial conditions at the other locations.

To investigate the general behavior of the model, we have obtained the temporal development of the concentration profiles at the mid point of the domain ($x = 0.5$m) for various parameter combinations of b and σ^2. The same realization of standard Wiener process increments and constant mean velocity of 0.5 m/day were used for all the computational experiments.

First we will illustrate that the SSTM can mimic the solution of advection-dispersion equation. We have used the concentration profiles given by the stochastic model as the observations of solute concentration to estimate the appropriate hydrodynamic dispersion coefficient (D) of the advection-dispersion model by using a stochastic inverse method, which will be described in Chapter 10. Using the inverse method we can answer the question: what would be the approximate dispersion coefficient given the concentration profiles from the SSTM assuming that the deterministic advection-dispersion equation can be applied? In this way we can compare the SSTM with the solutions of the advection-dispersion equation for the same boundary and initial conditions. The parameters of the SSTM, $\sigma^2 = 0.001$ and $b = 0.0001$, gave the corresponding estimate of 0.01 m^2/day for D and the SSTM can represent the advection-dispersion model with the estimated D (Figure 8.4).

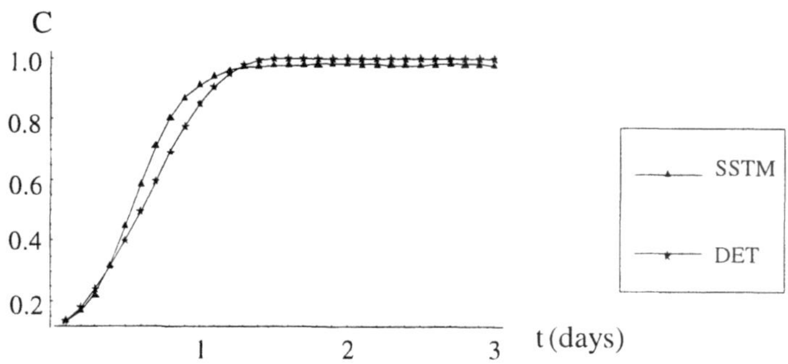

Figure 8.4 Comparison of deterministic (D = 0.01) and stochastic (σ^2 = 0.001 and b = 0.0001) model concentration profiles for 1 m domain.

Figure 8.5 illustrates that the SSTM could mimic the advection-dispersion model even for a larger scale, $0 \leq x \leq 10$ m, for the 30 day time period used in the calculation. We used the same SSTM parameters that were used in 1 m case and obtained the estimate of 0.037 m^2/day for D.

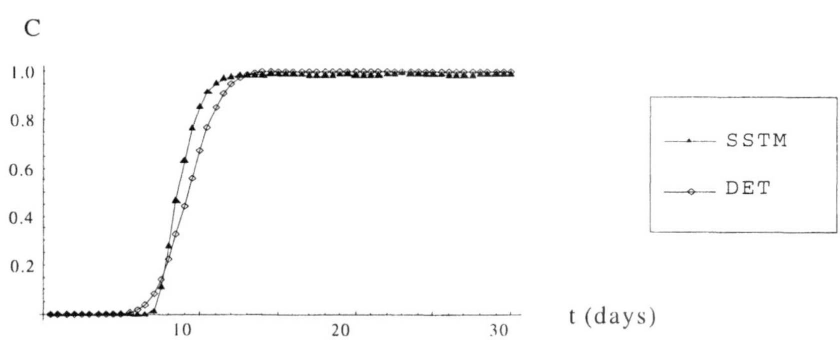

Figure 8.5 Comparison of deterministic (D = 0.037 m^2/day) and stochastic (σ^2 = 0.001 and b = 0.0001) model concentration profiles for 10 m domain.

We explored the changes of the statistical nature of the solutions with different b and σ^2. The behavioral change of the concentration breakthrough curves was examined by keeping one parameter at a constant and changing the other. Figure 8.6 shows the concentration profiles at x = 0.5 m of 1 m domain, for a smaller value of σ^2 (0.0001) when b varies from 0.0001 m to 0.25 m. The randomness of almost all five curves are insignificant, i.e. it is difficult to distinguish the different profiles. Although, range of b varies from 0.0001 to 0.25m (a change of 2500 times) the change of stochasticity is negligible for smaller σ^2. When σ^2 approaches 0, flow is advective and the dispersion is negligible.

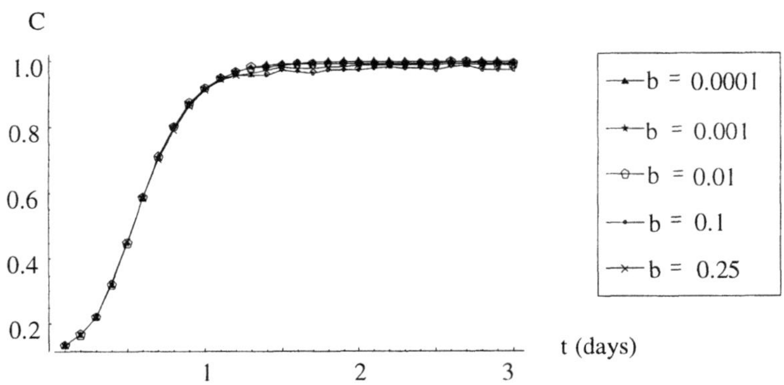

Figure 8.6 Concentration profile at x = 0.5m for σ^2 = 0.0001.

With the increase of σ^2 by 10 times for the same regime of b (0.0001 m to 0.25 m), Figure 8.7 shows visually distinguishable differences between concentration breakthrough curves. Furthermore, we can observe some curves have undergone notable stochasticity, especially when b = 0.1 m. The high values of variance not only directly increase the stochastic nature of the flow but also influence the ways in which b affects the flow. Another observation we can make from Figure 8.7 is that with the increase of stochasticity the concentration profile reaches its asymptotic value (sill) early and the maximum concentration value is less than those of more deterministic profiles.

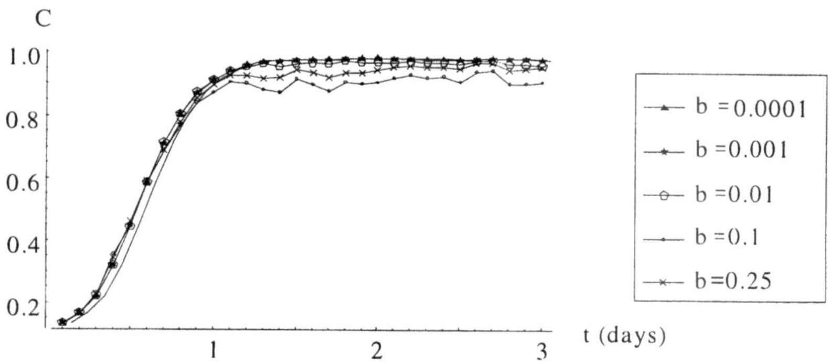

Figure 8.7 Concentration profile at $x = 0.5$ m for $\sigma^2 = 0.001$.

One can expect to see the increase of stochasticity with the increase of correlation length. When b is very small, flow is smooth and stable. However, it is interesting to see that, b at 0.1 m makes the concentration profile more variable. When b at a higher regime, 0.25m for example, the flow is less stochastic than it was at 0.1m. This may be caused by a sensitive range of b around 0.01 m. Figure 8.8 shows the concentration breakthrough curves for the similar b ranges at $\sigma^2 = 0.01$. Flow tends to be unsteady for larger correlation lengths; however, stochasticity of smaller b values is still trivial. Increase of σ^2 intensifies stochasticity and effect of b in the flow a great deal. The unpredictable behavior of the flow around 0.01 m of b shown in Figure 8.7 exists in current σ^2 as well.

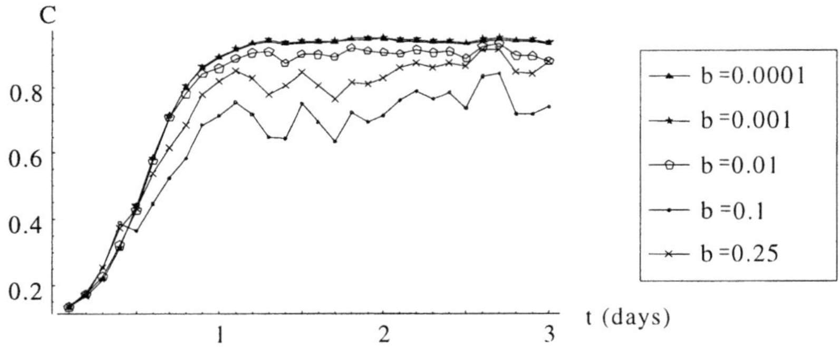

Figure 8.8 Concentration profile at $x = 0.5$ m for $\sigma^2 = 0.01$.

We extended the investigation by keeping b at a constant and changing σ^2. Figure 8.9 shows the concentration profiles at $b = 0.0001$ for varying σ^2 (0.0001 to 0.25). In Figure 8.6, small σ^2 demonstrates negligible stochasticity even for very high b values, whereas, in Figure 8.9, irrespective of smaller b, σ^2 influences the stochasticity of the flow. However, it is difficult to distinguish the concentration profiles for smaller σ^2 (0.0001 and 0.001). With the increase of σ^2 stochasticity increases rapidly. Therefore, we can assume that σ^2 is the dominant parameter which regulates the behavior of the flow.

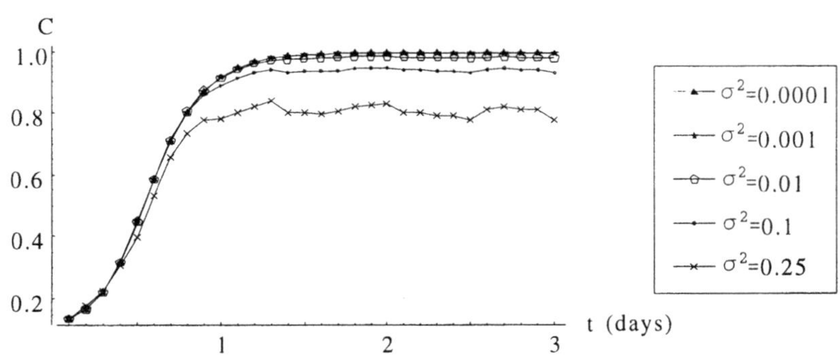

Figure 8.9 Concentration profile at $x = 0.5$m for $b = 0.0001$.

We increased b by 10 times and obtained Figure 8.10 which shows that randomness increases considerably.

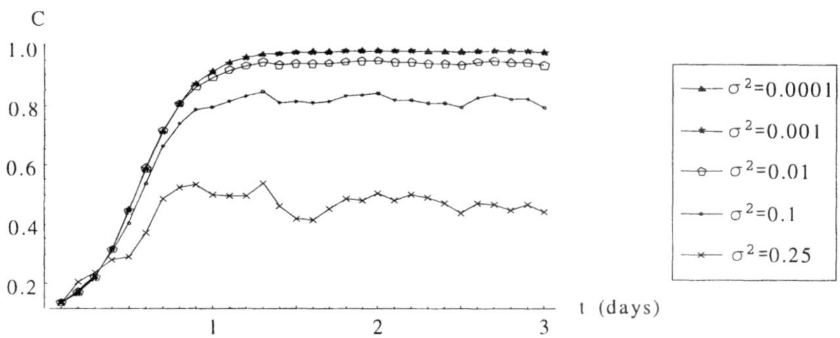

Figure 8.10. Concentration profile at $x = 0.5$m for b = 0.001.

It may be more appropriate and statistically sound to use confidence intervals rather than depend on a few realizations of the standard Wiener increments to understand the effect of σ^2. We used 50 different Wiener increments to calculate the 95% confidence intervals. Figure 8.11(a) clearly shows that for the smaller values of parameters (σ^2=0.001, b= 0.01), which represent less heterogeneity of the system, the variations of concentration profiles are negligible and hardly distinguishable. Figure 8.11(b) exhibits that when the parameter values are increased the stochasticity increases. The confidence intervals of Figure 8.11(b) demonstrate that the model is quite stable even for highly stochastic flow.

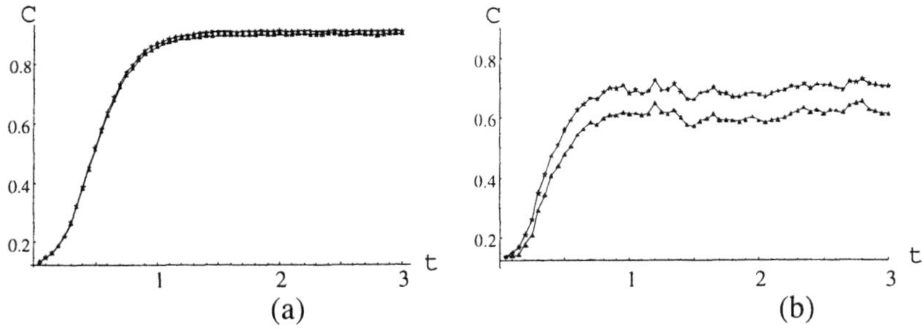

Figure 8.11 95% of confidence interval profiles with 50 different Wiener increments; (a) σ^2=0.001, b=0.01 ; (b) σ^2=0.1, b= 0.1.

We explored the effects of different random Wiener process increments. Figures 8.12 and 8.13 show the concentration profiles for five different Wiener processes for two different combinations of parameters. There are no considerable differences among these breakthrough curves, i.e. the influence of the Wiener process is minimal to the nature of the flow.

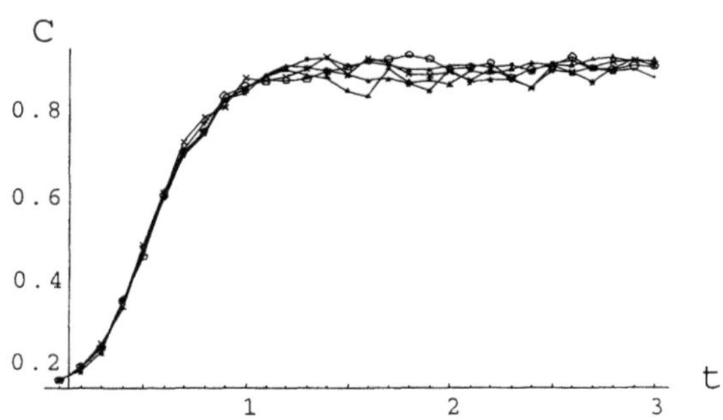

Figure 8.12 Concentration profiles for five different Wiener process increments at $x = 0.5$ m for $\sigma^2 = 0.001$ and $b = 0.01$.

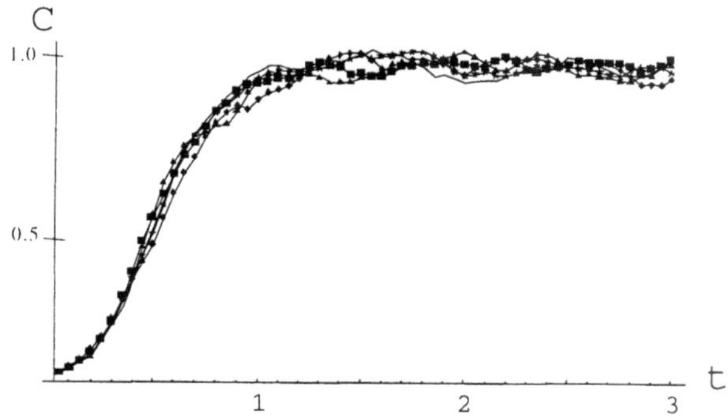

Figure 8.13 Concentration profiles for five different Wiener process increments at $x = 0.5$ m for $\sigma^2 = 0.01$ and $b = 0.1$.

8.6 Hypotheses Related to Variance and Correlation Length

Having understood some of the features of the model behavior, we can develop hypotheses about the parameters of the stochastic model relating to the physical phenomenon. As a fairly simple but reasonable attempt, it was hypothesized that the variance, σ^2, is a function of pore size and inversely proportional to the porosity ($\sigma^2 \propto (1/\varphi)$, where $\varphi =$ porosity). Low σ^2 represents larger pore sizes and more possible travel paths, i.e. solute can travel in water with fewer disturbances in less heterogeneous media. As a result, randomness of the travel paths and the occurrence of random mixing decrease. On other hand, larger σ^2 represents a medium of smaller pore size. Therefore, there are less straight travel paths and water tends to travel in various directions. This phenomenon can increase the mixing of the solute and, hence, increases dispersion and stochasticity.

We further hypothesized that the correlation length, b, is representative of the geometry of the pores. The small b represents the medium of isotropic and homogeneous formation, and larger b represents anistropic and heterogeneous porous medium.

When the pore sizes are fairly large the effect of the geometry is negligible. Flow paths can find easier ways through larger pores irrespective of the shapes of particles. Figure 8.6 shows that hypotheses are reasonable. Low σ^2,

0.0001, represents large pores, therefore the flow is stable for all the shapes of geometry (any b value). In the case of larger σ^2, where pore size is smaller, the geometry can play a vital role. Even though the effective pore size is smaller, if geometry of the pores are regular, particles could find a reasonably homogeneous paths and that comparatively reduces the random mixing of flow (Figure 8.7 and 8.8). In addition, the geometry and size of pores are interrelated in a complex manner.

We have also investigated the effect of parameters for the larger scales: 10 m, 20 m, 30 m, 50 m and 100 m (see Figures 8.14 to 8.17). Figure 8.14 and Figure 8.15 show that increase of stochasticity with σ^2 for 10 m domain. Comparison of Figure 8.14 and Figure 8.16 illustrate that σ^2 is the most dominant parameter and our hypotheses seem to be reasonable for larger scales as well.

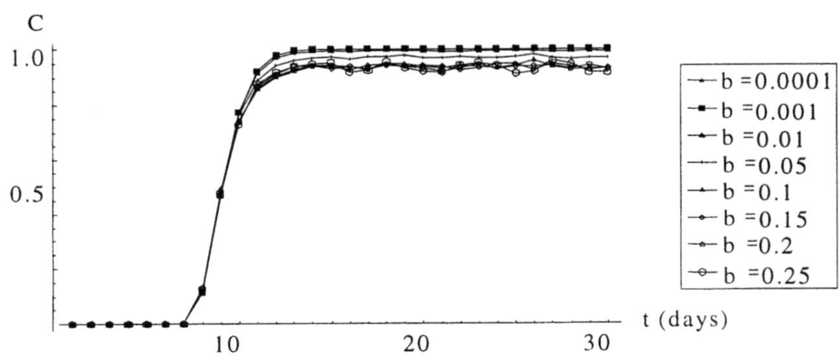

Figure 8.14 Concentration profile at x = 5m (of 10 m domain) for σ^2 = 0.0001.

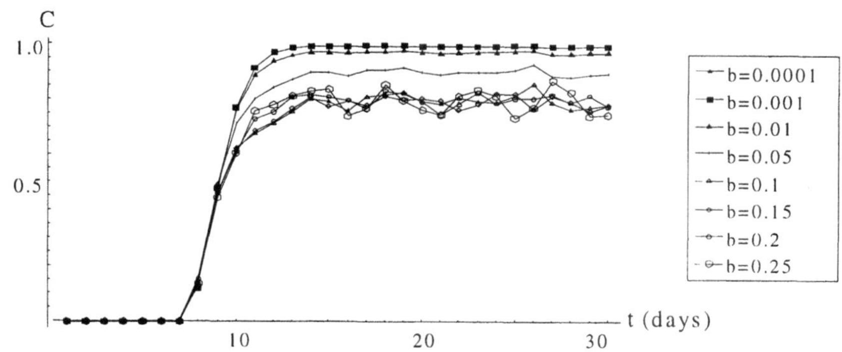

Figure 8.15 Concentration profile at x = 5m (of 10 m domain) for $\sigma^2 = 0.001$.

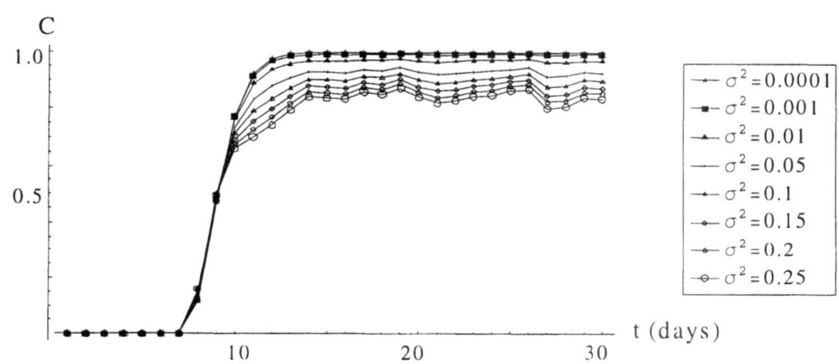

Figure 8.16 Concentration profile at x = 5m (of 10 m domain) for b = 0.0001.

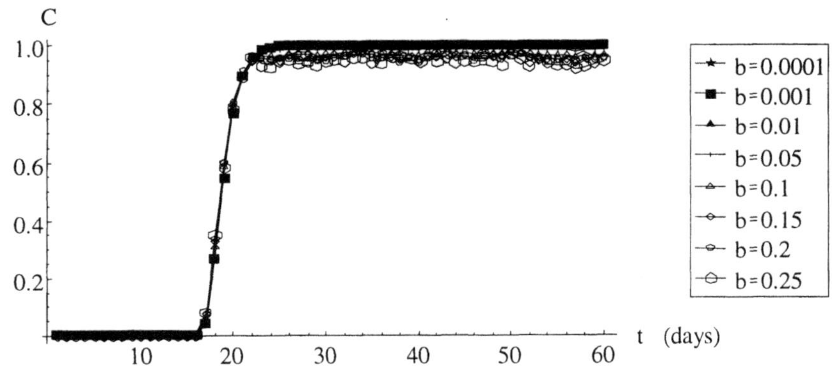

Figure 8.17 Concentration profile at x = 10m (of 20 m domain) for $\sigma^2 = 0.0001$.

8.7 Scale Dependency

As we have discussed in Chapter 1, the scale dependency of the hydrodynamic dispersion coefficient is an outcome of the associated Fickian assumptions. In developing the SSTM, we have not used any linearizing assumptions, but we have assumed a covariance kernel based on plausible arguments. It is reasonable to ask the question whether the parameters in the SSTM are dependent on the scale of the experiment. As the covariance kernel is based on the measures related to the properties of porous medium that are intrinsically affecting the flow, such as geometry of particle and pore sizes, we could expect low level of scale dependency in the case of the SSTM.

Comparison of Figure 8.6 and Figure 8.14 shows that stochasticity of the flow has increased with the scale of the experiment for the same parameters. Figures 8.7, 8.15, and 8.16 illustrate the same. Even though similar model performances are evident in other scales, visual comparison may not be sufficient to conclusively support capturing of scale dependency.

Can we observe the scale dependency of D by taking the concentration profiles from the SSTM as experimental observations and estimating the dispersion coefficient arising from the deterministic advection-dispersion equation? We employed the stochastic inverse method mentioned earlier to estimate D by using concentration realizations of the SSTM. As Figure 8.4 and Figure 8.5 show, D has increased from 0.01 m^2/day to 0.037 m^2/day with the scale for same parameters. However, as will be shown later, the reliability of the estimates obtained from the stochastic estimation method reduces with the increase in stochasticity. Therefore, such estimation method may not be suitable to estimate parameters with highly stochastic flows where values of σ^2 and b are large. However, flow with low level stochasticity illustrates that the SSTM is capable of explaining the scale dependency of D, but this needs to be explored in more detail.

8.8 Validation of One Dimensional SSTM

The validation of a model is the process of comparing the appropriate outputs of a model with the corresponding observations of the real system. It is a common practice to compare the model performance against available accepted models, which we did by comparing the model with widely accepted and commonly used advection-dispersion model. Our investigation showed that the SSTM could mimic the deterministic advection-dispersion model with reasonable accuracy at the 1-D level. In this section, we compare the results of the SSTM with those of the contaminant transport tests conducted at large, confined, artificial aquifers at Lincoln University, Canterbury, New Zealand. Furthermore, we use this exercise as a primary step for understanding the issues relating to the aspect of practical application of the SSTM.

8.8.1 Lincoln University Experimental Aquifers

Lincoln University experimental aquifers (2 in number) are 9.49 m long, 4.66 m wide and 2.6 m deep. As shown in Figure 8.18 constant head tanks bound the aquifer at its upstream and downstream ends. A porous wall provides the hydraulic connection between the aquifer and head tanks. A weir controls the water surface elevation in each head tank, and each weir can be adjusted

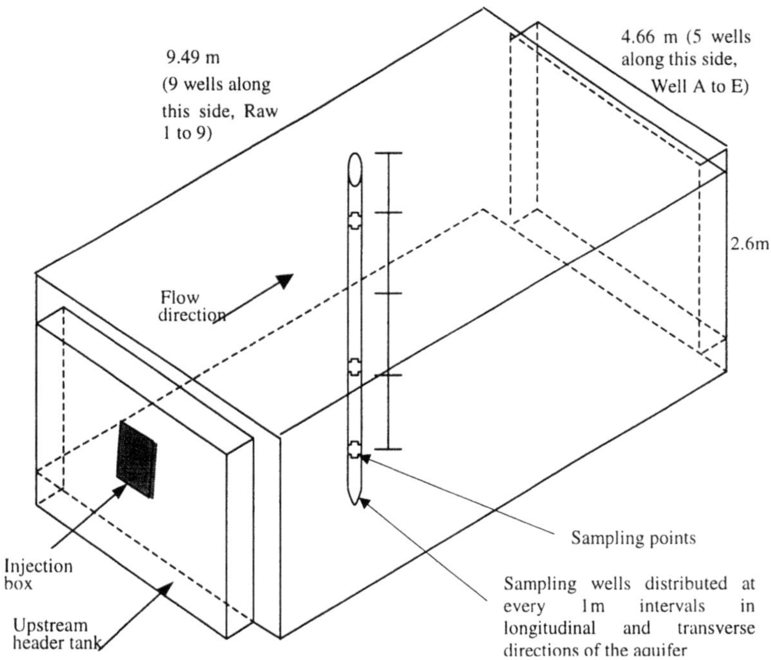

Figure 8.18 Schematic diagram of artificial aquifer at Lincoln University, New Zealand (Courtesy of Dr. John Bright, Lincoln Ventures Ltd.)

to provide different hydraulic gradients. However, the uniform hydraulic gradient of 0.017m/9.49m (= 0.0018) was maintained during the entire experiment. All other boundaries are zero flow boundaries. The aquifer media is sand.

Multi-port monitoring wells are laid out on a 1m x 1m grid. Computer controlled peristaltic pumps enable fully automated, simultaneous solute water samples to be collected from sample points that are uniformly distributed throughout the aquifer (four sample points for each grid point at 0.4m, 1.0m, 1.6m and 2.2m depth from the top surface of the aquifer). The tracer used was Rhodamine WT (RWT) dye with an initial concentration of 200 parts per billion and then allowed to decrease exponentially. Tracer was injected at middle of the header tank by using an injection box (dimensions of 50 cm length, 10 cm width and 20 cm depth). This tracer was rapidly mixed into the upstream header tank and thus infiltrated across the whole of the upstream face of the aquifer. The dye was injected at 12.00 noon and samples were collected at 2 to 4 hour intervals (however, there are some exceptions on time intervals) for 432 hours.

8.8.2 Methodology of Validation

Since, the present stochastic model is one-dimensional, we experimented with directly relating to solute concentration profiles in a single dimension of the aquifer. However, as one can assume, the actual aquifer is subjected transverse dispersion and consideration of mere one-dimensional flow is not sufficiently accurate. Therefore, we employed the following methodology to approximate the aquifer parameters.

There are solute concentration values of the artificial aquifer available for a large number of spatial points for different temporal intervals. Mainly the data are available for header tank, row 1, row 3, row 5, row 7 and row 9 (see Figure 8.18) for all the levels. First we selected few spatial coordinates at row 5 of well A – level YE. Then, we developed a two-dimensional deterministic advection-dispersion transport model and obtained corresponding concentration values from the model at the selected spatial locations of the aquifer. As the past studies show, we assumed that transverse dispersion coefficient is 10% of the longitudinal dispersion (Fetter, 1999). The mean velocity of 0.5 m/day was considered. Afterwards the profiles of both the aquifer and the deterministic model were plotted in one axis system to compare their similarity. This curve fitting technique was carried out in association with trial and error exercises to determine a most suitable fitting of the curves by changing dispersion coefficient of the deterministic model.

8.8.3 Results

After investigating many combinations of parameters by trial and error, we found that the longitudinal dispersion coefficient at closer to 0.1 m^2/day is giving a reasonable fit (Figure 8.19). Following figures show a sample of results. We found that closest fit is given by longitudinal dispersion coefficient of 0.15 m^2/day, and the transverse dispersion is 0.015 m^2/day. For the simplicity, the concentration values of the aquifer were normalized.

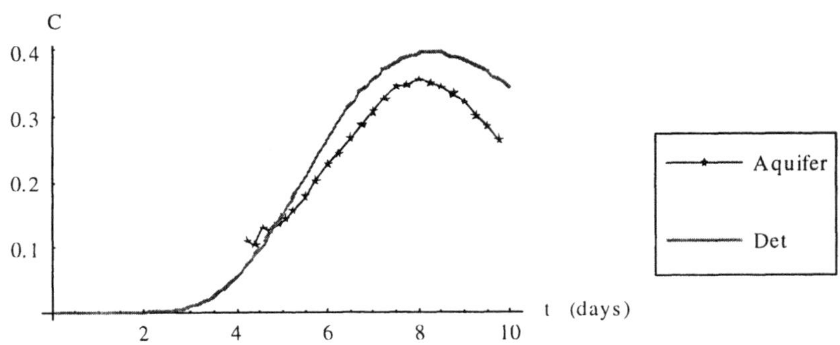

Figure 8.19 Concentration profile of trial and error curve fit for longitudinal dispersion coefficient of 0.1 m^2/day of advection dispersion model with row 5 of aquifer data.

However, the longitudinal dispersion coefficient of 0.15 m^2/day provides the best fit (see Figure 8.20).

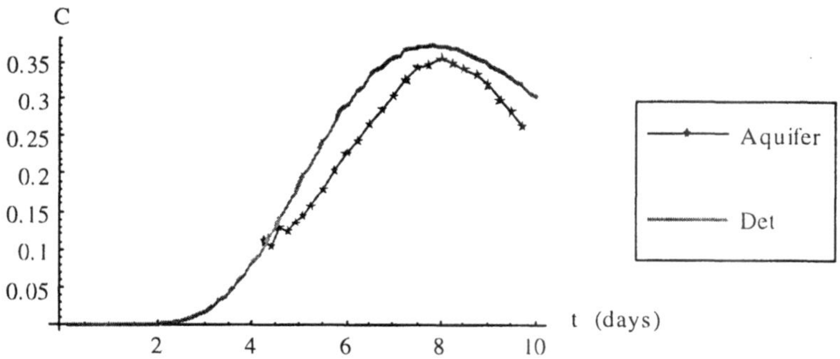

Figure 8.20 Concentration profile of trial and error curve fit for longitudinal dispersion coefficient of 0.15 m^2/day of advection dispersion model with row 5 of aquifer data.

Subsequently we developed a one-dimensional deterministic advection-dispersion model by using the longitudinal dispersion coefficient obtained from two-dimensional comparisons. As mentioned above, such a coefficient may be a more realistic representation of the artificial aquifer data transformed to 1-D. Then we used the same curve fitting technique that was used above, with the 1-D deterministic model and the 1-D stochastic model (SSTM). Investigation of curve fitting for different parameter combinations was conducted for the same Wiener process. The following are a sample of our trial and error curve fittings (Figures 8.21 - 8.24).

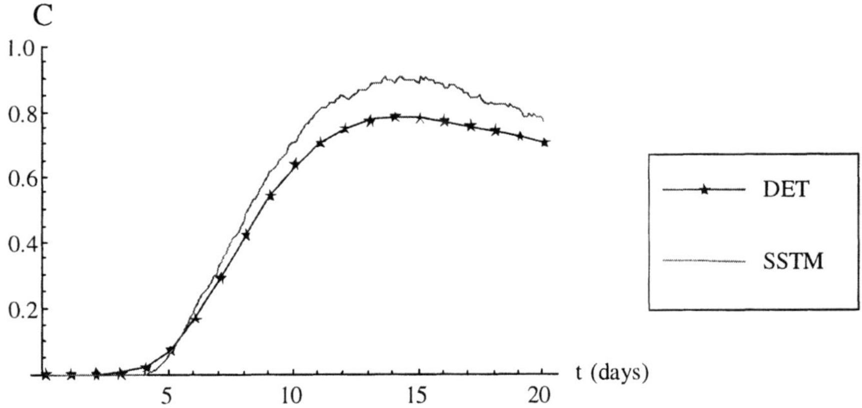

Figure 8.21 Concentration profiles of deterministic advection-dispersion model ($D = 0.15$ m^2/day) and SSTM with $\sigma^2 = 0.001$ and $b = 0.01$.

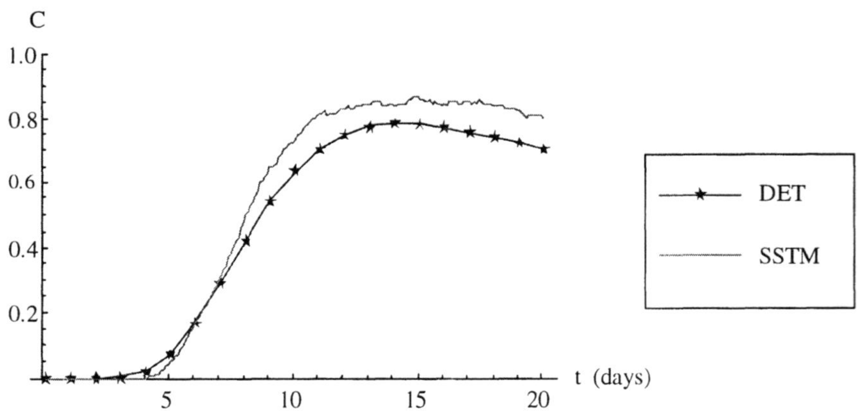

Figure 8.22 Concentration profiles of deterministic advection-dispersion model ($D = 0.15$ m^2/day) and SSTM with $\sigma^2 = 0.01$ and $b = 0.001$.

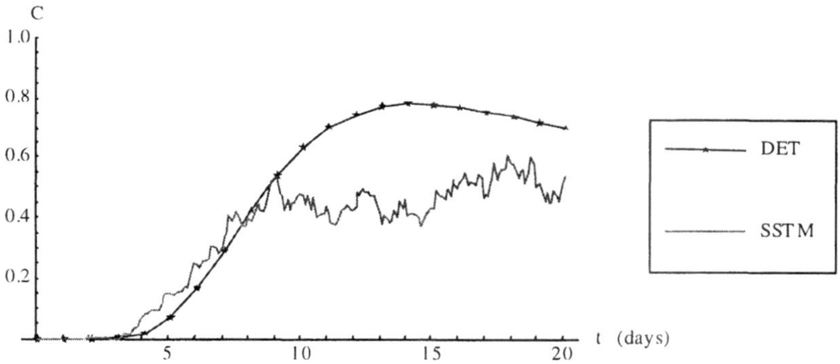

Figure 8.23 Concentration profiles of deterministic advection-dispersion model (D = 0.15 m^2/day) and SSTM with $\sigma^2 = 0.1$ and $b = 0.01$.

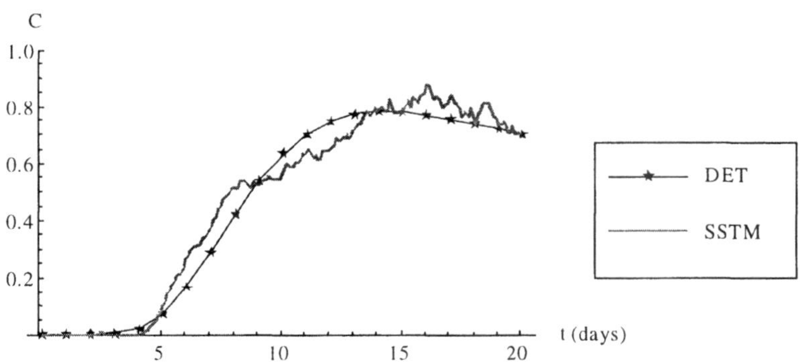

Figure 8.24 Concentration profiles of deterministic advection-dispersion model (D = 0.15 m^2/day) and SSTM with $\sigma^2 = 0.01$ and $b = 0.01$.

As figures 8.21 – 8.24 show parameter combination of the stochastic model that closely represent the aquifer data were $\sigma^2 = 0.01$ and $b = 0.01$.

Having determined the appropriate parameters of the SSTM that simulates the Lincoln University aquifer at a selected spatial location (Row 5 – well A) we investigated the robustness of the model for different Wiener processes. Figure 8.25 shows that model is reasonably stable for seven different sets of Wiener increments.

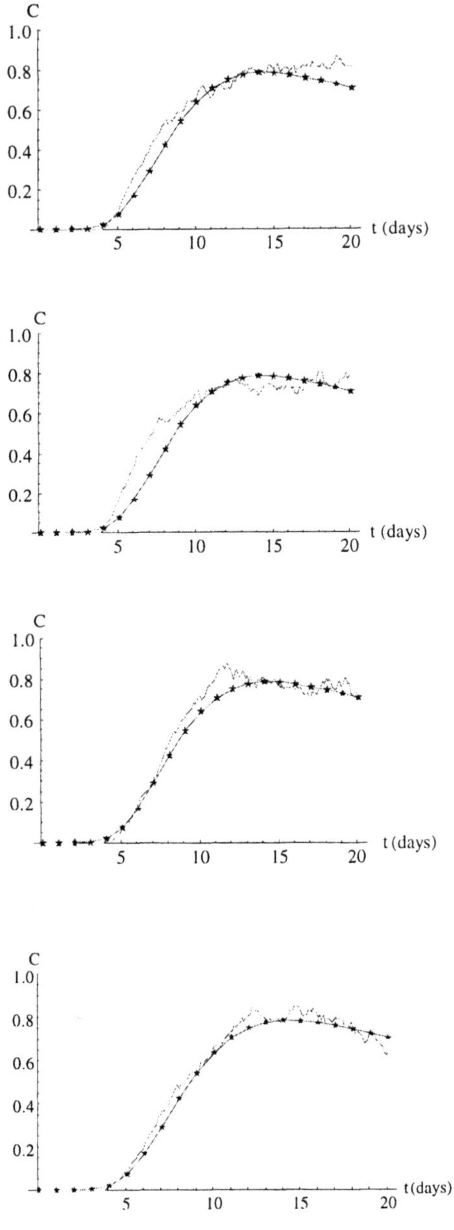
C
1.0
0.8
0.6
0.4
0.2
5
10
15
20
t (days)
C
1.0
0.8
0.6
0.4
0.2
5
10
15
20
t (days)
C
1.0
0.8
0.6
0.4
0.2
5
10
15
20
t (days)
C
1.0
0.8
0.6
0.4
0.2
5
10
15
20
t(days)

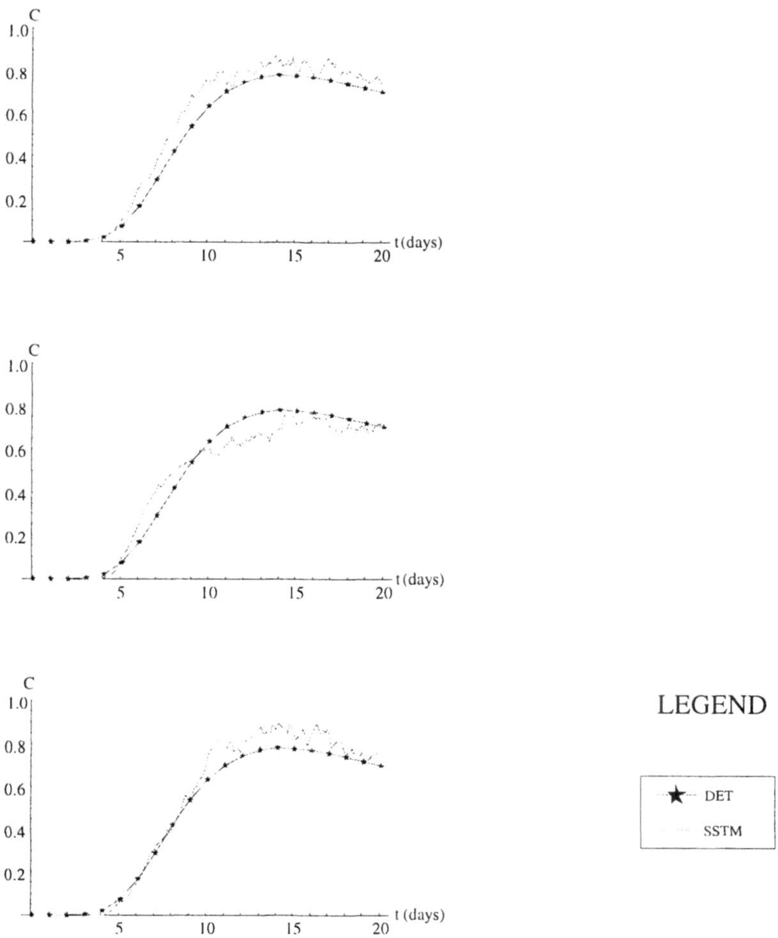

Figure 8.25 Concentration profiles of deterministic advection-dispersion model (D = 0.15 m^2/day) and SSTM with $\sigma^2 = 0.01$ and $b = 0.01$ for seven different standard Wiener processes.

Even though the above-mentioned results shows that parameter combination of $\sigma^2 = 0.01$ and $b = 0.01$ is a fairly accurate representation of the experimental aquifer for the given spatial point, we extended the validation process for other spatial locations. Concentration data of row 3 of well A was considered. The methodology described in the previous section was applied for this data set as well. Figure 8.26 shows that 2-D deterministic advection-dispersion model with longitudinal dispersion coefficient of 0.15 m^2/day reasonably fit the aquifer data for similar location.

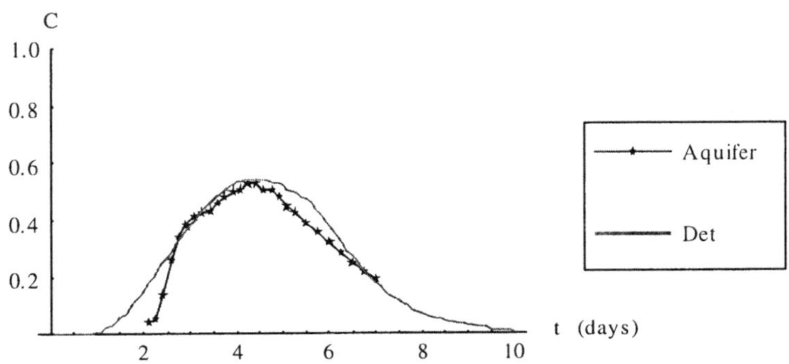

Figure 8.26 Concentration profiles of 2D deterministic advection-dispersion model (D = 0.15 m^2/day) and the experimental aquifer.

Since the given longitudinal dispersion coefficient is a plausible representation of the new spatial location, we continue the process into the next step to compare the curves of one-dimensional deterministic advection-dispersion model and those of the SSTM. We used the same parameters, σ^2= 0.01 and b = 0.01, utilized in the above section for the present spatial point as well. Figure 8.27 shows that the curves of the deterministic model and SSTM are in a reasonable agreement.

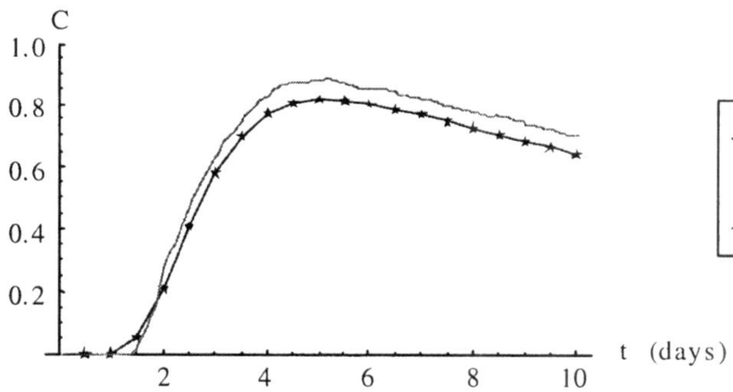

Figure 8.27 Concentration profiles of deterministic advection-dispersion model ($D = 0.15$ m^2/day and SSTM with $\sigma^2 = 0.001$ and $b = 0.01$ for row 3 well A.

We extended our comparison to row 7 of the same well. Figure 8.28 and Figure 8.29 show that same parameter combination is reasonably valid for this spatial point as well.

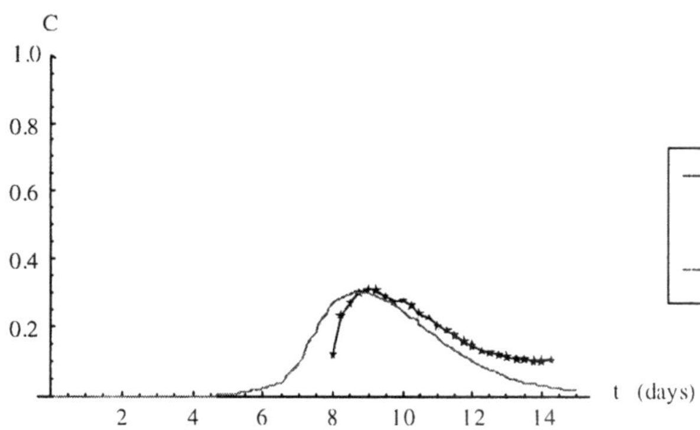

Figure 8.28 Concentration profiles of 2D deterministic advection-dispersion model ($D = 0.15$ m^2/day) and the experimental aquifer at row 7.

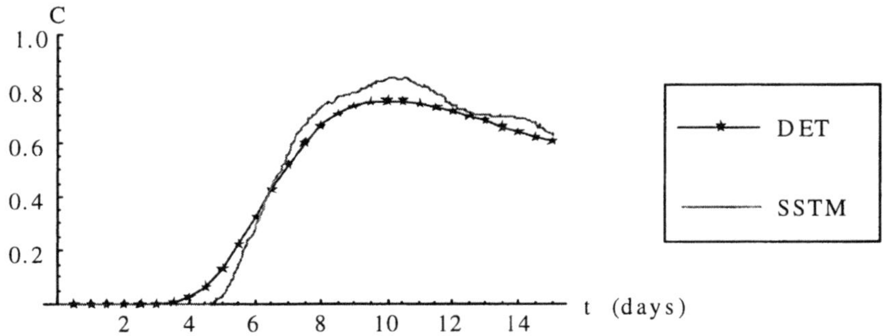

Figure 8.29 Concentration profiles of deterministic advection-dispersion model (D = 0.15 m^2/day) and SSTM with $\sigma^2 = 0.001$ and $b = 0.01$ for row 7 well A.

8.7 Concluding Remarks

In this chapter we have developed a solute transport model using the concepts in stochastic calculus and tested the model using computations and comparing the outputs with the measured data from the artificial aquifer experiments. The new approach used to develop a stochastic solute transport model can be used to mathematically describe the dispersion without resorting to the Fickian assumptions. Even with a simpler covariance kernel for the velocity in 1-D, SSTM can produce satisfactory results with respect to the data collected from the aquifer. We have also showed that the parameters in the model could be related to the real aquifer properties but there is a lot to be done in that direction. Further research also needs to be done to characterize different kernels and the associated parameters based on currently available data from the experimental aquifers and natural underground formations.

Chapter 9

Solving the Eigenvalue Problem for a Covariance Kernel with Variable Correlation Length

9.1 Introduction

In the previous chapter, we have shown that a stochastic model of flow in a porous medium (e.g. in an aquifer) can be formulated in terms of a flow equation that is perturbed by a spatially distributed Wiener process. The tortuous trajectory of a fluid element is modeled as a response to random fluctuations in medium properties such as porosity and hydraulic conductivity that it encounters as it moves through the medium. Thus the medium properties can be considered to vary smoothly on a macroscopic scale, but to have random fluctuations over a microscopic scale superimposed on it, reflecting the granularity of the medium. The scale of this granularity appears in the mathematical description as a spatial correlation between the stochastic perturbations at nearby points; i.e. a correlation length b appears in the description as an important parameter describing medium properties.

The way in which the correlation length enters the stochastic differential equation (SDE), is contained in a Karhunen-Loeve expansion of the Wiener term (Ghanem and Spanos, 1991). In the Karhunen-Loeve expansion, independent Wiener processes at different points are replaced by a single process modulated by a function of position. It turns out that if this modulation is constructed from the eigenfunctions of some assumed covariance function q(x,y), the stochastic variation of the Wiener processes are indeed correlated as specified by q within a range determined by the correlation length that appears as a parameter in q.

By an eigenfunction of q, we mean a function $f_n(x)$ that satisfies the following integral equation:

$$\int_p^q q(x_1, x_2) f_n(x_2) dx_2 = \lambda_n f_n(x_1). \qquad (9.1)$$

Here, p and q are constant limits describing an interval of interest, e.g. the boundaries of the flow region. For each value of index n, a discrete function is to be found satisfying this equation for a corresponding real number λ_n, the eigenvalue. It remains to choose a covariance function. In this article, we only consider a function of the form:

$$q(x_1, x_2) = e^{-\frac{|x_1 - x_2|}{b(x_1, x_2)}} . \tag{9.2}$$

In terms of integral equation terminology, the covariance function plays the role of a kernel in equation (9.1) The solution of this equation for an exponential kernel of the form of equation (9.2) but with a constant value of b, is well known (Ghanem and Spanos, 1991). Reducing the integral equation to a differential equation yields a function of the form:

$$f_n(x) = A_n \cos(\omega_n \tfrac{x}{t}) + B_n \sin(\omega_n \tfrac{x}{t}) . \tag{9.3}$$

As equation (9.3) indicates, it is convenient to rescale all spatial variables, including b, to dimensionless variables by dividing by the interval length t = q-p. The coefficients A_n and B_n are determined by boundary conditions at x = p and x = q that are dictated by the integral equation. This, in turn, only has non-trivial solutions for the discrete values ω_n that satisfy the equation:

$$\tan \omega = \frac{2b\omega}{b^2\omega^2 - 1} . \tag{9.4}$$

The eigenvalues of the integral equation are related to the ω_n by

$$\lambda_n = \frac{2b}{b^2\omega_n^{\ 2} + 1} \tag{9.5}$$

In the present context, the outlined solution would describe a homogeneous medium. However, the purpose of this article is to extend that to the case that b is also a function of position, in order to apply the stochastic model to non-homogeneous media. The λ_n and f_n are the quantities needed for expanding spatially varying Wiener process amplitudes, and as shown elsewhere (Ghanem and Spanos, 1991) the covariance function itself can also be expanded in terms of them:

$$q(x_1, x_2) = \sum_n \lambda_n f_n(x_1) f_n(x_2) . \tag{9.6}$$

To set the scene for the various approximations to be discussed, it is useful to establish some plausible properties of the covariance function. Firstly, the Wiener process is perfectly self-correlated, i.e. q(x,y) = 1 for x = y (a condition relaxed for a general Wiener process), and the correlation must decrease as |x-y| increases. This means that q(x,y) is a peaked function along the line x = y in the X-Y plane. Equation (9.2) clearly exhibits this behavior. Secondly, it may be assumed that the peak width is small compared to the dimensions of the interval of interest, i.e. the (scaled) b(x,y) << 1. This is because the correlation length is much smaller than the macroscopic dimensions of the aquifer.

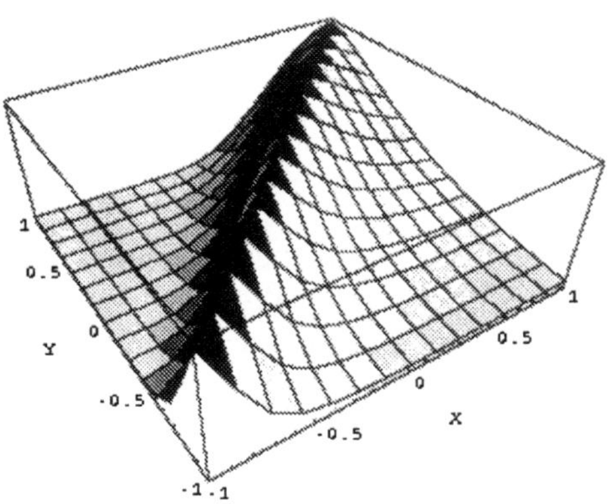

Figure 9.1 Schematic view of a variable correlation length covariance function; b increases in the direction away from the viewer.

Finally, the fact that changes in b are to be used to model changes in the medium properties from one part of the aquifer to another, is plausibly represented by the assumption b(x,y)=b((x+y)/2), i.e since only nearby points are significantly correlated, the correlation length may be evaluated at the midpoint between them. These characterizations are summarized by the schematic plot of a plausible covariance function in Figure 9.1.

Section 9.2 will be devoted to an outline of various proposed methods to solve the variable b eigenvalue problem. Section 9.3 compares the accuracy and computational effort when these are applied to a specific example of an assumed position dependence of b. We discuss the conclusions that can be drawn from the comparison in section 9.4.

9.2 Approximate Solutions

Equation (9.1) is classified as a Fredholm integral equation of the second kind (Morse and Feshbach, 1953). The exact solution for constant b discussed above was obtained by applying the standard technique to reduce an equation of this kind to a differential equation. However, when b is variable, this does not deliver a differential equation that is easily solved, and moreover in the applications envisaged b may only be known as a table of numerical values derived from measured media properties. Hence analytical methods are ruled out, and we resort to numerical solutions.

There are also well documented standard techniques for numerical solution of Fredholm equations of the second kind (Press et al., 1992). Prominent among these is the Nystrom method, which uses Gauss-Legendre integration on the kernel integral to reduce the integral equation to a matrix eigenvalue problem of dimension equal to the number of integration points. To evaluate the method, it was applied to equation (9.1) for a fixed value b = 0.2 for which the analytical solution is known. Comparing the eigenvalues found with the exact ones, improvements were found up to about 40 integration points, after which numerical inaccuracies set in. Eigenvalues could be obtained to within 10%, but the eigenfunctions are highly irregular and do not resemble the smooth exact functions given by equation (9.3). The reason for this failure is that the simple Nystrom method only works well for a smooth kernel. The exponential kernel however, is nearly singular - while it does remain finite, its derivative across the diagonal line x = y is discontinuous and it is highly localized around this line.

For the treatment of a kernel with a diagonal singularity, the Nystrom method is often extended by making use of the smoothness of the solution to subtract out the singularity (Press et al, 1992). When applied to the present case, this is found to give some improvement for a low number of integration points but it is actually worse for more than about 12 points. The best accuracy obtained is no better than for the simple Nystrom method. More elaborate methods to deal with diagonal singularities have been used; for example, methods that construct purpose made integration grids to take the singular behavior into account (Press et al, 1992). However, we aim to construct a method which does not require a detailed prior knowledge of the kernel, and so these methods do not appear promising.

Moreover, if a specialized method is anyway required, a more direct approach is to make use of the known analytical solution for the fixed b case. This is supported by noting that the solutions in equations (9.2) – (9.5) do not, in fact, depend strongly on the value of b. That is illustrated by Figure 9.2, which shows the behavior of the n=4 eigenfunction for 0.001<=b<=0.5, a variation over more than 2 orders of magnitude.

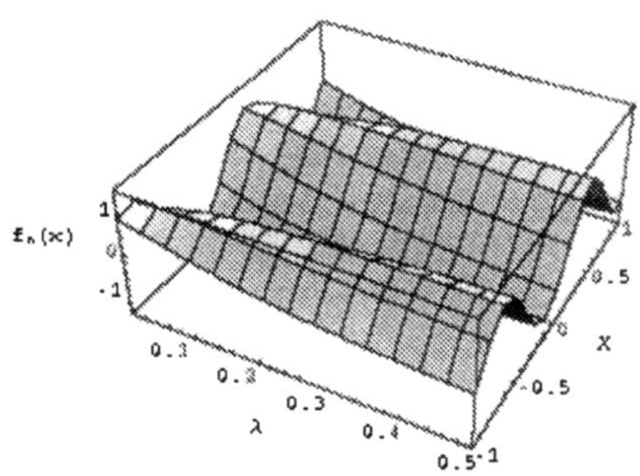

Figure 9.2 The n=4 eigenfunction of a fixed correlation length kernel, as the constant value b=λ, ranges from λ = 0.001 to λ = 0.5.

An obvious way to exploit this observation, is to expand the eigenfunctions for variable b in terms of those calculated for some fixed typical correlation length b^0, e.g. the average value of b(x,y) over the integration interval:

$$f_n(x) = \sum_m F_{nm} f_m^{\,0}(x)\,. \tag{9.7}$$

When this is substituted into equation (9.1), the integral eigenvalue equation for the function q(x,y) is transformed to a matrix eigenvalue equation for the matrix Q defined by:

$$Q_{nm} = \int_p^q dx_1 \int_p^q dx_2 q(x_1, x_2) f^0{}_n(x_1) f^0{}_m(x_2)\,. \tag{9.8}$$

The dimension of the matrix is equal to the cutoff value M that has to be introduced as upper limit of the expansion over m in equation (9.7). Once the matrix has been diagonalized, the elements F_{nm} of its eigenvector matrix can be substituted back into equation (9.7) to get the first M of the desired eigenfunctions and its eigenvalues are identical to the first M eigenvalues of the integral equation.

The eigenfunctions of the kernel with a fixed correlation length b^0 can be shown to form a complete orthogonal basis. Therefore this method to solve the variable b case is exact up to the introduction of the finite cutoff M. Because the eigenfunctions are relatively insensitive to the value of b it is reasonable to expect a fast convergence of the expansion, so for practical purposes it should be possible to keep M fairly small. Nevertheless this solution is computationally intensive, not only because each of the M^2 elements of Q requires a multiple integral, but because the near singularity in q requires a large number of integration points for accurate numerical integration. The approximate methods described below are intended to overcome this problem.

A key observation in this regard is that the double integration in equation (9.8) can be reduced to a single integral if b is a constant. By splitting the inner integral into two subranges the absolute value in the exponent in q can be eliminated, and in each subrange a factor $\exp(\pm x_1/b)$ can be factored out of the integral provided that b does not depend on x_2. The remaining integrand can be analytically integrated because of the simple form of the f^0_n as specified by equation (9.3), leaving only the outer integral to be done numerically.

A direct way to take advantage of this idea is to approximate $b(x_1,x_2)$ as piecewise constant. The behavior of $q(x_1,x_2)$ limits significant contributions to the integral to the vicinity of the diagonal line $x_1 = x_2$. Thus in a subdivision of the region of integration into a grid of square blocks, the dominating contribution will come from those blocks strung along the diagonal. In each of these q is approximated by using a fixed value of b, e.g its value in the centre of the block. The matrix element integral is reduced to a sum of integrals over the diagonal blocks, in each of which a different constant value of b is used to reduce it to a one-dimensional integral. We refer to this as the piecewise kernel matrix (PKM) method.

Having decided to use a piecewise kernel, one can go a step further by also constructing piecewise eigenfunctions. In fact, in this framework it is plausible to do away with the matrix problem altogether. Since a formula for the eigenfunction corresponding to any one of the piecewise constant values of b is known, this solution may be used within the subinterval, and the complete eigenfunction constructed by linking up all the solutions across the subinterval boundaries. The comparison between this approach and the matrix approach is somewhat like that between a spline function interpolation and a Fourier expansion of a function. However, in the present context the eigenfunctions to be linked up are already largely determined and there are not enough free parameters available to ensure that the function and its derivative are continuous across the subinterval boundary (as is done by spline functions). In fact, a problem in applying piecewise eigenfunctions is to determine the relative amplitudes of the functions used in neighboring subintervals. As the eigenvalue equation is independent of amplitude, the only guideline is the overall normalization over the entire interval. In practice, the insensitivity of the eigenfunctions to b ensures that discontinuities remain insignificant if subintervals are chosen to allow only moderate change of b from one subinterval to the next.

The elimination of the need to calculate and diagonalize a matrix in the piecewise eigenfunction (PE) method, is a major conceptual simplification. However, in computational terms it is not so much simpler. If there are M subintervals, for each eigenfunction M sets of coefficients in each subinterval need to be kept, and that is similar to keeping coefficients for an expansion over M basis functions in a matrix method. Also, all subsequent manipulations with piecewise eigenfunctions require the complexity of breaking up operations into subintervals, while in the matrix method a single function valid over the whole interval is obtained even when it was calculated from a piecewise kernel.

Returning to the matrix methods, there is another way to obtain the benefits of a constant b in calculating the matrix element integral. We write

$$b(x_1, x_2) \approx b(\tfrac{x_1+x_2}{2}) \approx b(x_1) \tag{9.9}$$

i.e. the correlation length b is kept variable, but only its value on the diagonal is used, because the behavior of q limits the effective region of integration to $x_1 \approx x_2$. Equation (9.9) is enough to allow the factorization of the kernel that leads to one-dimensional matrix element integrals. This is described as the diagonal correlation length matrix (DCLM) method.

9.3 Results

To evaluate the relative performance of the various methods, we take p = -1, q = 1 and

$$b(x) = 0.1 + 0.05x\,. \tag{9.10}$$

Thus b varies by a factor of three over the interval, but is always small compared to the total interval size t=2. None of the methods make any use of the assumed functional dependence; only numerical values are used, either pointwise or per subinterval. Therefore the simple linear variation of b should be representative of any other smooth variation within a similar range of values.

As a first measure of performance, it is noted that equation (9.1) can be interpreted as an integral operator that when operating on an eigenfunction, gives back the same function apart from an amplitude factor which is the corresponding eigenvalue. Therefore a candidate eigenfunction can be judged by how closely it resembles the resultant function obtained from it by the action of the integral operator. Figure 9.3 shows the n=4 eigenfunction and its (rescaled) resultant, for the 3 approximations.

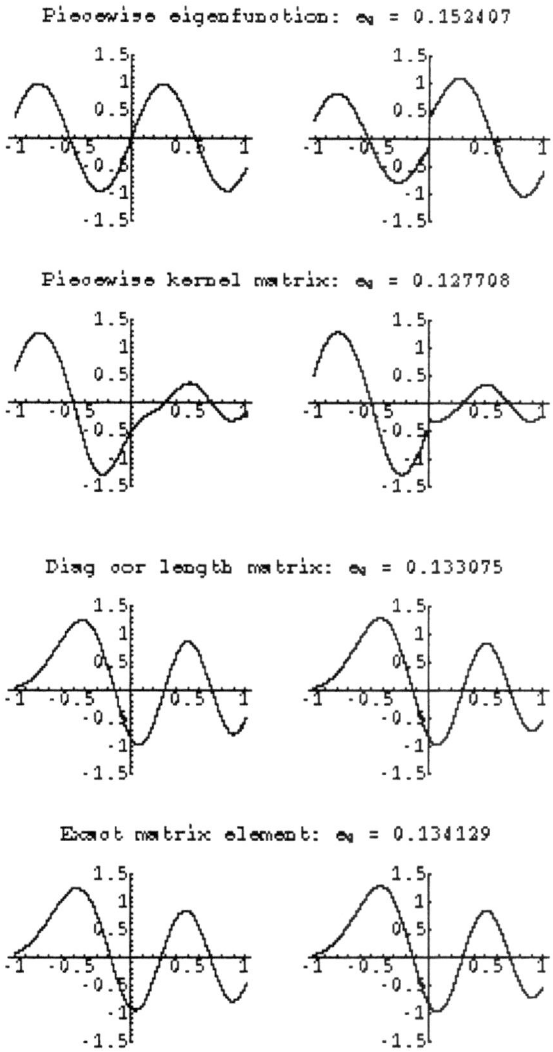

Figure 9.3 Eigenfunctions (left) and resultant (right) from action of kernel operator.

A very coarse piecewise grid consisting of only two subintervals (-1,0) and (0,1) was chosen for the piecewise approximations shown in the first two sets of graphs in figure 9.3, while a dimension of M=9 eigenfunctions was used for all the matrix methods. The calculations have also been repeated for a piecewise grid of 9 subintervals (which makes the complexity comparable to that of the 9x9 matrix methods) but despite the smaller step in kernel values from one subinterval to the next, the accuracy is worse because for the particular choice of parameters in equation (9.10) the assumption b<< subinterval is not satisfied for 9 subintervals.

The calculated eigenfunctions in Figure 7.3 all seem to satisfy the requirement that they should remain unchanged by the action of the kernel integral, reasonably well. Nevertheless the functions look quite different, apart from a qualitative agreement in the number of peaks, and also the eigenvalues differ substantially.

An obvious defect of the piecewise eigenfunction is that its amplitude in the two subintervals is the same whereas all the others show a plausible response to the different value of the kernel in the subintervals. The PE method might be improved if a rule can be found to choose relative amplitudes in the subintervals, but that would clearly introduce a discontinuity into the function.

The PKM method on the other hand appears to exaggerate the amplitude difference. By contrast, the DCLM approximation gives an eigenfunction that is virtually indistinguishable from the exact one in the last row of Figure 9.3, and also shows an excellent agreement of the eigenvalue.

A more quantitative measure of the performance of the eigenfunctions can be constructed by making use of the fact that the matrix elements of the kernel operator, defined as in equation (9.8) but using the calculated variable b eigenfunctions, should be the diagonal eigenvalue matrix. For each of the approximate methods, this 9x9 matrix was calculated and each element Q_{nm} scaled by dividing it by the harmonic mean of the diagonal elements at positions n and m. This produces a matrix with unit elements on the diagonal and off-diagonal elements should be 0 on a scale of 0 to 1. The largest scaled off-diagonal element is used as a measure of the overall eigenvector accuracy. Similarly, the eigenvalue accuracy is measured by the maximum difference from the exact value scaled to a fraction of the exact value. Unlike the visual comparison presented above, these measures cover the complete set of eigenfunctions and eigenvalues that were calculated. The results of the comparison is shown in Table 9.1, along with a comparison of the computing times for the methods.

Table 9.1. Time and accuracy comparison for various approximation methods.

Approx	CPU time	Eigenval dev, %	Eigenfun dev, %
Exact	4400	-	-
PE, 2 sub	27	21	89
PE, 9 sub	116	47	77
PKM, 2 sub	9	8	14
PKM, 9 sub	40	36	13
DCLM	20	3	3

It is seen that all the approximate methods are faster than an exact calculation by about two orders of magnitude, but only the DCLM method achieves this without a substantial loss of accuracy, particularly regarding the eigenfunctions. The "exact" calculation used as reference, uses matrix elements that are calculated without approximation to within the numerical accuracy of the underlying quadrature routines, but is itself approximate to the extent that the matrix is truncated to 9x9. However, to confirm that this is not in practice a significant limitation, the DCLM calculation was repeated with the cutoff increased to a dimension of M=20, and agreement was found within the convergence limits of the quadrature routines for the first 9 eigenvalues and -functions.

The purpose for calculating the eigenfunctions in the context of this article is to be used to expand stochastic amplitudes. Therefore it is also of interest to see how they perform as basis functions for expanding a known function; the obvious candidate is the covariance function itself, for which the expansion is known to be given by equation (9.6). Figure 9.4 shows the result of that comparison.

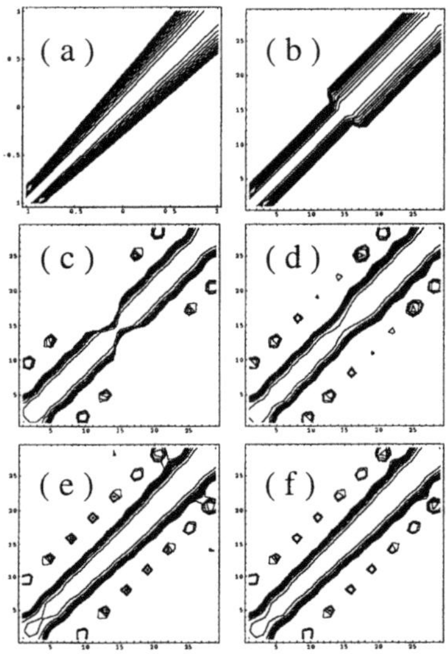

Figure 9.4 Contour plots of various representations of the exponential covariance function.

a) Exact kernel	b) stepwise kernel
c) 9-term PE expansion	d) 9-term PKM exp.
e) 9-term DCLM exp.	f) 9-term exact exp.

Again it is seen that the DCLM expansion gives the best representation of the covariance function, hardly distinguishable from that given by the exact eigenfunctions. Even the latter has some difficulty in representing the exponential rise in value near the diagonal because of the truncation of the expansion. However, it should be noted that for the sake of clarity, the contour spacing in the diagrams is logarithmic rather than equally spaced, so that the outlying contours showing the most deviation represent very small function values.

9.4 Conclusions

Of the various approximations investigated, the DCLM method gives the most accurate eigenfunctions and eigenvalues, virtually identical with the ones calculated from exact matrix elements but requiring less computing time by two orders of magnitude.

The actual calculations reported here, were done using the Mathematica® on a 150 MHz Pentium computer. The absolute timing values can no doubt be improved on by faster machines and using compiled code, but the relative values should remain similar.

It is concluded that the DCLM is a practical method for solving the eigenvalue problem for an exponential covariance kernel with a correlation length that is position dependent, and the application of this method to stochastic modeling of flow problems in porous media will be further investigated.

Chapter 10

A Stochastic Inverse Method to Estimate Parameters in Groundwater Models

10.1 Introduction

Natural systems are heterogeneous and they contain noise due to random inputs, irregular varying coefficients and fluctuations in boundary conditions. In this chapter, we model the behavior of natural systems using stochastic differential equations, present a parameter estimation procedure for such models in a general setting, and extend it to simple groundwater models. The applications to groundwater models are within the context of one dimensional solute transport problem to estimate parameters for two governing equations, one consisting of a single parameter and other of two parameters. The results of this inverse methodology are reliable in the presence of noise. However, the investigation of solute transport parameter estimates shows an inverse relationship to the noise level. The main advantage of the estimation methodology presented here is its direct dependence on field observations of state variables of natural systems in the presence of uncertainty.

As we have seen before, we can model the behavior of natural systems such as groundwater flow and solute transport in porous media through differential equations based on conservation laws. In the process of developing the differential equations, we introduce the parameters, which we consider attributes or properties of the system. In the case of groundwater flow, for example, the parameters such as hydraulic conductivity, transmissivity and porosity are constant within the differential equation, and it is often necessary to assign numerical values to these parameters. These values of the parameters are obtained from laboratory experiments and/or field scale experiments. However, these values may not represent the often complex patterns across a large geographic area, hence limiting the effectiveness of the model. In addition, such field scale experiments can be expensive. Often we are interested in modeling for quantities such as the depth of water table and solute concentration. This is because they are directly relevant to environmental decision making, and we measure these variables regularly and the measuring techniques tend to be cheaper. Further we can continuously monitor these decision (output) variables in many situations. Therefore it is reasonable to assume that these observations of the output variables represent current status of the system. If the dynamics of the system can reliably be modeled by a relevant differential equation, we can expect the parameters estimated based on the observations may give us more reliable representative values than those obtained from laboratory tests and literature. However, such observations often contain noise from two different sources: experimental errors and noisy system dynamics. Noise in the system dynamics may be due to heterogeneity of the media, random nature of inputs such as rainfall and variable boundary conditions to name a few factors. The question of estimating the parameters from the observations naturally involves the models that represent the system noise as well. In this Chapter, we aim to illustrate a parameter estimation procedure for such models containing noise. We present models in a general setting so that the procedure can be used for groundwater models. Then we apply parameter estimation theory and procedures to the solute transport in saturated porous media in the presence of noise.

10.2 System Dynamics with Noise

Let us consider a differential equation of the form,

$$\frac{dy}{dt} = f(\theta, \; y, \; t) \tag{10.1}$$

where y is the dependent variable (output) that is observed, t is time and θ is a parameter upon which the characteristics of the model depends. Suppose we can include the noise ($\xi(t)$) contained in dy/dt as an additive component to equation (10.1). For simplicity we will assume that ξ depends only on time. In many engineering and natural systems, this noise is irregular, continuous and independent of each other, and white noise has been considered as a valid approximation (Øksendal, 1998). Therefore, we can express equation (10.1) as

$$\frac{dy}{dt} = f(\theta, \; y, \; t) \; + \; \xi(t) \; , \tag{10.2}$$

and multiplying by dt we get

$$dy = f(\theta,\ y,\ t)dt + \xi(t)dt\,. \tag{10.3}$$

Once we consider white noise as a model for the noise term, y becomes a stochastic process having many realizations or paths. A set of observations of y can be considered as a realization of y. Considering that the noise represents deviations away from the deterministic rate, the expected value of $\xi(t)$ over all realizations is zero; $\xi(t)$ is an independent stochastic process; the joint probability distribution of $\xi(t)$ is time-invariant; and $\xi(t)$ has to be continuous though irregular. The only stochastic process that can meet all these requirements is the Wiener process $(B(t))$ based on observations of the Brownian motion (Øksendal, 1998). It can be shown that,

$$dB(t) = \xi(t)dt\,, \tag{10.4}$$

taking the convergence in the probability.

In equation (10.4), $dB(t)$ are increments of the standard Wiener process which are normally distributed with a unit variance (for a detailed discussion refer to the chapters 2, 3 and 4).

Substituting in equation (10.3) we have,

$$dy = f(\theta,\ y,\ t)dt + dB(t)\,;\quad 0 \le t \le T\,. \tag{10.5}$$

This is a stochastic differential equation giving the drift term ($f(\theta,\ y,\ t)$) and diffusive term ($dB(t)$). Kutoyants (1984), for example, gives the likelihood function $L(\theta)$ to estimate θ given the observation for y under certain conditions:

$$L(\theta) = \exp\left\{\int_0^T f(\theta,\ y,\ t)dy(t) - \frac{1}{2}\int_0^T f^2(\theta,\ y,\ t)dt\right\}. \tag{10.6}$$

Taking natural logarithm of both sides, the log likelihood is given by

$$l(\theta) = \int_0^T f(\theta,\ y,\ t)dy(t) - \frac{1}{2}\int_0^T f^2(\theta,\ y,\ t)dt\,. \tag{10.7}$$

By maximizing $l(\theta)$ with respect to θ we obtain,

$$\frac{\partial l(\theta)}{\partial \theta} = \int_0^T \frac{\partial}{\partial \theta}(f(\theta, y,t))dy(t) - \int_0^T f(\theta, y,t) \frac{\partial}{\partial \theta}(f(\theta, y,t))dt = 0. \tag{10.8}$$

Equation (10.8) will give the maximum likelihood parameter $(\hat{\theta})$ for θ, given the values of y. We will illustrate the use of equation (10.8) by taking an example.

10.2.1 An Example

Suppose that the dynamics of a system could be expressed by,

$$\frac{dX(t)}{dt} = \theta \ X(t) + \xi(t); \quad X(0) = 1, \quad 0 \le t \le 1 \tag{10.9}$$

where $X(t)$ is the process under observation, θ is a parameter to be determined from the observations, and $\xi(t)$ is the noise component assumed to be white. Following the arguments mentioned in the previous section, we can express the process $X(t)$ in terms of a stochastic differential:

$$dX(t) = \theta \ X(t)dt + dB(t). \tag{10.10}$$

Following the definition of Ito integral (Øksendal, 1998) we can explicitly solve equation (10.10) and the solution can be expressed in terms of Wiener process, $B(t)$:

$$X(t) = X(0) \ \exp(B(t))\exp((\theta - 1/2)t) \tag{10.11}$$

where $B(t)$ is the standard Wiener process. The solution of equation (10.11) consists of a set of realizations of $X(t)$; and as an example, a realization of $X(t)$ is given in Figure 10. 1 for $\theta = 1.5$. Let us assume that we observe the realization of $X(t)$ depicted in Figure 10.1 and we seek to estimate θ given $X(t)$ and corresponding time values.

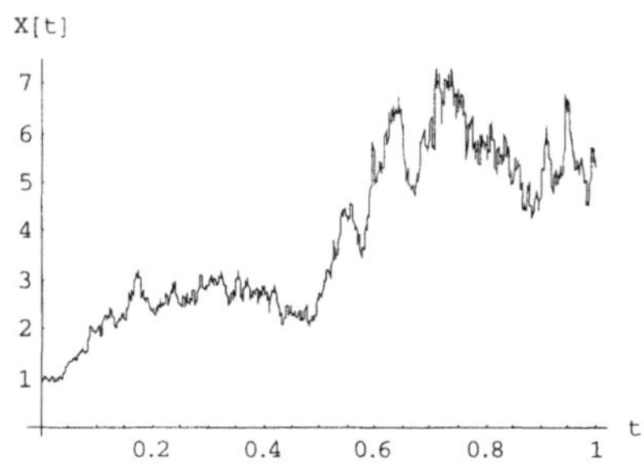

Figure 10.1 A realization of equation (10.11) for θ =1.5.

In this case, comparing to equation (10.5)

$$f(\theta, y, t) = \theta \; X(t)$$

and equation (10.8) can be expressed as

$$\frac{\partial l(\theta)}{\partial \theta} = \int_0^1 X(t)dX(t) - \int_0^1 \theta X(t)X(t)dt,$$

$$= \int_0^1 X(t)dX(t) - \int_0^1 \theta X^2(t)dt.$$

To maximize the log likelihood,

$$\frac{\partial l(\theta)}{\partial \theta} = 0.$$

Therefore, $$\hat{\theta} = \frac{\int_0^T X(t)dX(t)}{\int_0^T X^2(t)dt}. \qquad (10.12)$$

From the given observations (Figure 10.1) we express $\hat{\theta}$ as

$$\hat{\theta} \approx \frac{\sum_{i=1}^{M} X(t_i) \quad \Delta X(t_i)}{\sum_{i=1}^{M} X^2(t_i) \quad \Delta t}. \tag{10.13}$$

The estimated value of θ $(\hat{\theta})$ for the given realization is 1.47849. Similarly, to investigate the robustness of the procedure, we compute $\hat{\theta}$ from 30 different realizations of $X(t)$. Calculated from equation (10.11) with $\theta = 1.5$, the mean of $\hat{\theta}$ is 1.48135 with a standard deviation of 0.586255. This shows that if we sample the $X(t)$ process from equation (10.11) a reasonable number of times, the mean value of $\hat{\theta}$ is very close to 1.5.

It would be interesting to see what happens when the standard Wiener increment term ($dB(t)$) in equation (10.5) is modified by an amplitude (σ^2):

$$dy = f(\theta, \ y, \ t)dt + \sigma^2 \ dB(t). \tag{10.14}$$

Equation (10.12) is used to calculate θ, and Table 10.1 shows the mean and the standard deviations of $\hat{\theta}$ based on 30 distinct realizations of $X(t)$ from equation (11) with $\theta = 1.5$. It is evident that an amplitude of 1.0 or less, and slightly above 1.0 would produce reliable estimates from this procedure.

Table 10.1 Mean and standard deviations of parameter estimates.

	$\hat{\theta}$	
σ^2	Mean	Std. Deviation
0.01	1.47319	0.117916
0.10	1.55374	0.234585
0.25	1.39394	0.385624
0.50	1.52886	0.426577
1.00	1.48135	0.586255
1.50	1.76876	1.424700

10.3 Applications in Groundwater Models

In this section, the above described general parameter estimation procedure is applied in the context of solute transport in saturated porous media in the presence of noise. Unny (1989) presented the basis for this application by describing groundwater system in the form of stochastic partial differential equations and then estimating parameters.

10.3.1 Estimation Related to One Parameter Case

The stochastic one-dimensional advective transport equation can be expressed as,

$$\frac{\partial C}{\partial t} = -v_x \left(\frac{\partial C}{\partial x}\right) + \xi(x,t) \tag{10.15}$$

where

v_x = average linear velocity, m/day,
C = solute concentration, mg/l,
and $\xi(x, t)$ is described by a zero-mean stochastic process.

We multiply equation (10.15) by dt throughout and as in equation (10.14), formally replace $\xi(x, t)dt$ by $\sigma^2 dB(t)$ (see Unny (1989) for the derivation). Now, we can obtain the stochastic partial differential equation as follows,

$$\partial C = -v_x \left(\frac{\partial C}{\partial x}\right) dt + \sigma^2 dB(t). \tag{10.16}$$

Suppose we have observations of solute concentration, C_i at M independent space coordinates along x-axis, where $1 \leq i \leq M$, at different time intervals, t (where $0 \leq t \leq T$). In other words we have M number of C_i observations for each time step. Hence, altogether, there are $((T+1)*M)$ number of C_i observations. We use these observations to estimate the parameter θ, which, in this case v_x (or hydraulic conductivity, if hydraulic gradient and porosity are known), of all possible parameter values using maximum likelihood approach. As we explained above, we can write equation (10.16) in the form of equation (10.14),where

$$f(\theta, C, t) = -v_x \left(\frac{\partial C}{\partial x}\right).$$

The likelihood expression for the estimation of parameter θ can be given by

$$L(\theta) = \exp\left\{\int_0^T f(\theta, C, t)\, dC(t) - \frac{1}{2}\int_0^T f^2(\theta, C, t) dt\right\}. \tag{10.17}$$

The estimate $\hat{\theta}$ can be obtained by maximizing $L(\theta)$; therefore,

$$\frac{\partial L(\theta)}{\partial \theta} = 0. \tag{10.18}$$

If, $l(\theta) = \ln L(\theta)$, (10.19)

taking the natural log on both sides of the equation (10.17)

$$l(\theta) = \int_0^T f(\theta, C, t)\, dC(t) - \frac{1}{2}\int_0^T f^2(\theta, C, t) dt. \tag{10.20}$$

The parameter is estimated as the solution to the equation

$$\int_0^T \frac{\partial}{\partial\theta} f(\theta, C, t)\, dC(t) - \int_0^T f(\theta, C, t) \frac{\partial}{\partial\theta} f(\theta, C, t) dt = 0. \tag{10.21}$$

If we observe M independent sample paths, the likelihood-function becomes the product of the likelihood functions for M individual sample paths,

$$L(\theta) = L(\theta, C_1)\, L(\theta, C_2) \ldots\ldots L(\theta, C_M). \tag{10.22}$$

Taking the natural log on both sides of the equation (10.22) we have the log-likelihood,

$$l(\theta) = l(\theta, C_1) + l(\theta, C_2) + \ldots\ldots + l(\theta, C_M). \tag{10.23}$$

Therefore, the log likelihood function can be expressed as

$$l(\theta) = \sum_{i=1}^{M} \int_0^T f(\theta, C_i, t)\, dC_i(t) - \frac{1}{2} \sum_{i=1}^{M} \int_0^T f^2(\theta, C_i, t) dt, \tag{10.24}$$

and the parameter estimate $\hat{\theta}$ is obtained as the solution to maximum likelihood

$$\sum_{i=1}^{M} \int_0^T \frac{\partial f(\theta, C_i, t)}{\partial\theta} dC_i(t) - \sum_{i=1}^{M} \int_0^T f(\theta, C_i, t) \frac{\partial f(\theta, C_i, t)}{\partial\theta} dt = 0. \tag{10.25}$$

Let us assume that the drift term in equation (10.14), $f(t, C, \theta)$, depends linearly on its parameters θ, then we can express it as

$$f(\theta, C, t) = a_0(C, t) + \theta\, a_1(C, t). \tag{10.26}$$

The log-likelihood function from equation (10.24) is

$$l(\theta_1) = \sum_{i=1}^{M} \int_0^T \{a_0(C_i,t) + \theta a_1(C_i,t)\} dC_i(t) - \frac{1}{2} \sum_{i=1}^{M} \int_0^T \{a_0(C_i,t) + \theta a_1(C_i,t)\}^2 \, dt. \quad (10.27)$$

The estimate $\hat{\theta}$ is obtained as a solution to the equation

$$\sum_{i=1}^{M} \int_0^T \{a_1(C_i,t)\} dC_i(t) - \sum_{i=1}^{M} \int_0^T \left\{a_0(C_i,t) + \theta a_1(C_i,t)\right\} \left\{a_1(C_i,t)\right\} dt = 0. \quad (10.28)$$

Hence the estimate of $\theta(\hat{\theta})$ is given by

$$\hat{\theta} = \frac{\sum_{i=1}^{M} \int_0^T \{a_1(C_i,t)\} dC_i(t) - \sum_{i=1}^{M} \int_0^T \{a_0(C_i,t)\} \{a_1(C_i,t)\} dt}{\sum_{i=1}^{M} \int_0^T \left\{a_1^2(C_i,t)\right\} dt}. \quad (10.29)$$

When we compare equations (10.16) and (10.26), we have

$$a_0(C_i,\ t) = 0; \quad a_1(C_i,\ t) = -\left(\frac{\partial C}{\partial x}\right)_i; \ \theta = v_x.$$

Therefore, the estimate of v_x is given by,

$$\hat{v}_x = \frac{\sum_{i=1}^{M} \int_0^T -\left(\frac{\partial C}{\partial x}\right) dC_i(t)}{\sum_{i=1}^{M} \int_0^T \left\{\left(\frac{\partial C}{\partial x}\right)_i\right\}^2 dt}. \quad (10.30)$$

10.3.2 Estimation Related to Two Parameter Case

We can use the same theoretical basis to estimate two parameter space problems. As an example, let us consider a one-dimensional stochastic advection-dispersion equation, which is given by

$$\frac{\partial C}{\partial t} = D_L \left(\frac{\partial^2 C}{\partial x^2}\right) - v_x \left(\frac{\partial C}{\partial x}\right) + \xi(x,t), \tag{10.31}$$

where D_L is the longitudinal hydrodynamic dispersion coefficient, m^2/day. Two parameters to be estimated are D_L and v_x. Equation (10.26) can be written in the following form:

$$f(t,C,\theta) = a_0(C,t) + \theta_1 a_1(C,t) + \theta_2 a_2(C,\ t). \tag{10.32}$$

In a similar way to the one parameter problem, we can compare equation (10.32) and the drift term of equation (10.31):

$$a_0(C_i,t) = 0; \quad a_1(C_i,t) = \left(\frac{\partial^2 C}{\partial x^2}\right)_i; \quad a_2(C_i,\ t) = -\left(\frac{\partial C}{\partial x}\right)_i; \theta_1 = D_L;$$
$$\theta_2 = v_x .$$

The log-likelihood function from equation (10.21) is

$$\begin{aligned} l(\theta_1,\theta_2) = & \sum_{i=1}^{M}\int_0^T \{a_0(C_i,t)+\theta_1 a_1(C_i,t)+\theta_2 a_2(C_i,t)\}\, dC_i(t) \\ & -\frac{1}{2}\sum_{i=1}^{M}\int_0^T \{a_0(C_i,t)+\theta_1 a_1(C_i,t)+\theta_2 a_2(C_i,t)\}^2\, dt. \end{aligned} \tag{10.33}$$

Differentiating (10.33) with respect to θ_1 and θ_2 respectively we get the following two simultaneous equations:

$$\sum_{i=1}^{M}\int_{0}^{T}\{a_1(C_i,t)\}dC_i(t)-\sum_{i=1}^{M}\int_{0}^{T}\{a_0(C_i,t)+\theta_1 a_1(C_i,t)+\theta_2 a_2(C_i,t)\}\{a_1(C_i,t)\}dt=0. \quad (10.34a)$$

$$\sum_{i=1}^{M}\int_{0}^{T}\{a_2(C_i,t)\}dC_i(t)-\sum_{i=1}^{M}\int_{0}^{T}\{a_0(C_i,t)+\theta_1 a_1(C_i,t)+\theta_2 a_2(C_i,t)\}\{a_2(C_i,t)\}dt=0. \quad (10.34b)$$

Now we obtain the values for $\hat{\theta}_1$ and $\hat{\theta}_2$ as the solutions to these two equations.

10.3.3 Investigation of the Methods

We use the above-mentioned method to estimate parameters in equations (10.30) and (10.34) by using a noisy dataset. The one dimensional solute transport dataset was generated by using equation (10.15) for one parameter case and equation (10.31) for the two parameter case. First, data was generated by using the deterministic solutions for each case and then noise was added randomly to each deterministic concentration value to generate a stochastic dataset. As an example, in the case of a maximum of ±5% introduced randomness, the noise component was generated by a random function which gives a maximum of 5% of deterministic concentration and another randomness function selects + or – operation. The spatial domain of the solution is 10m ($0 \le x \le 10$).

10.4 Results

The example in section 2.1 shows that the parameter estimation methodology described in this Chapter produces reliable estimates for a noisy dynamic system. The expected values of the estimates are closer to the actual parameters at low noise levels. As the percentage of noise is increased by changing the σ^2, the difference between actual and estimated parameters becomes larger. However, it is interesting to see that the mean value shows a close correlation to the actual value though the standard deviation increases with the noise (Table 10.1).

We present only a sample of results for the simulation study of solute transport. Figure 10.2 shows the estimated average linear velocity, v_x (0.3), that was used to generate the deterministic solution, against the actual parameter value for one parameter case. Figure 10.3 shows the comparison of the longitudinal dispersion coefficient, D_L in the two parameter estimation.

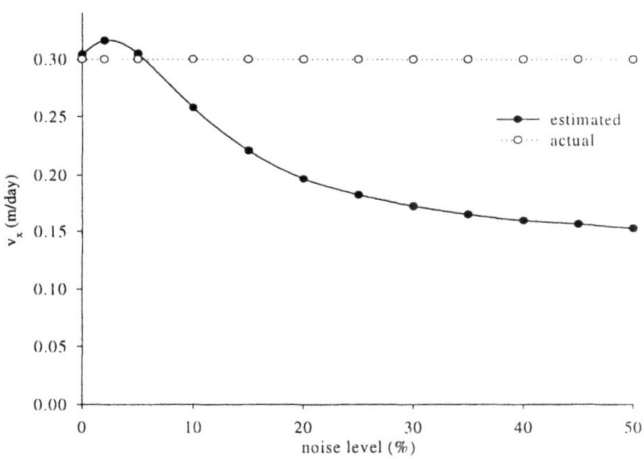

Figure 10.2 Actual and estimated velocity for different noise levels for one-parameter case.

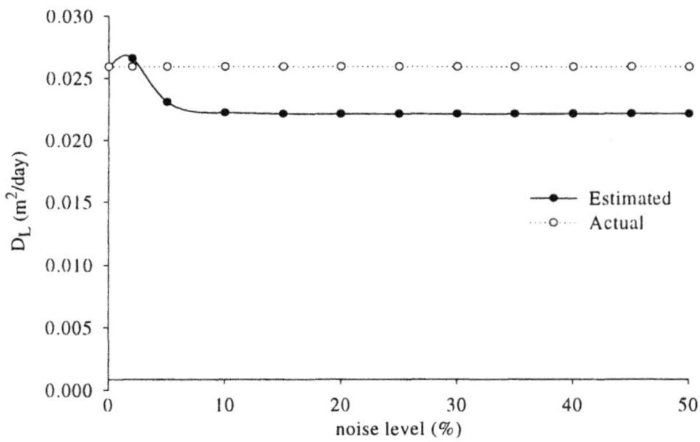

Figure 10.3 Actual and estimated longitudinal dispersion coefficient (D_L) for different noise levels in two-parameter case.

As seen in Figure 10.2 and Figure 10.3, the deviations of the estimated parameters from the corresponding actual values increase at first and then begin to flatten as the noise level increases. For example, the onset of flattening is 5% in Figure 10.2 whereas it is 2% in the Figure 10.3.

10.5 Concluding Remarks

In this Chapter, we have shown a straightforward procedure to estimate parameters of stochastic differential equations, which model the dynamics of systems containing noise. A sample of results has been discussed in two different cases to show that the likelihood functions give reasonable results even with significant levels of noise contained in the data. This procedure can be extended to the cases where the amplitude of noise is non-linear, but it is beyond the scope of this chapter.

References

Abramowitz and Stegun. 1965. Handbook of mathematical functions. Dover Publications, New York.

Anderson, M. 1979. Modeling of groundwater flow systems as they relate to the movement of contaminants. C.R.C. Crit. Rev. Environ. Control, (9), 97 - 156.

Bear, J. 1969. Hydrodynamic dispersion. IN: Flow through porous media. Academic Press, New York.

Bear, J. 1972. Dynamics of fluids in porous media. Elsevier, New York, NY.

Bear, J.; and D. K. Todd. 1960. The transition zone between fresh and salt waters in coastal aquifers. Univ. of California, Berkley, Hydraulics Lab, Water Resources Center, Contribution No. 29.

Bear, J.; and A. Veruijt. 1992. Modeling groundwater flow and pollution. Reidel Publishers, Holland.

Brown, T. N.; and D. Kulasiri. 1996. Validating models of complex, stochastic, biological systems. Ecological Modelling. (86), 129-134.

Chandrasekhar, S. 1943. Stochastic problems in physics and astronomy. Rev. Mod. Phys., (15), 1 - 87.

Crank, J. 1990. The mathematics of diffusion. Second Edition. Oxford University Press, UK.

Cushman, J.H. 1987. Development of stochastic partial differential equation in subsurface hydrology. J. Stach. Hydrol. Hydraul. (1), 241 - 262.

Dagan, G. 1988. Time-dependent macrodispersion for solute transport in anisotropic heterogeneous aquifers. Water Resour. Research, (24), 1491 - 1500.

Dagan, G. 1990. Transport in heterogeneous porous formation: spatial moments, ergodicity, and effective dispersion. Water Resour. Research, (26), 1281 - 1290.

Daniel, P. 1952. The measurement of groundwater flow. Ankara Symp. Arid Zone Hydrology, Proc. No. 2, UNESCO, 99 - 107.

Durrett, R. 1996. Stochastic calculus: a practical introduction. CRC.

Dynkin, E.B. 1965. Markov processes, vol I, Springer Verlag, Berlin.

Fetter, C.W. 1999. Contaminant hydrogeology, Second Edition, Prentice Hall.

Freyberg, D.L. 1986. A natural gradient experiment on solute transport in a sand aquifer, 2, Spatial movements and the advection and dispersion of nonreactive traces. Water Resour. Res., (22), 2031 - 2046.

Gelhar, L.J.; W. Mantoglou ; and K. R. Rehfeldt. 1985. A review of field-scale physical solute transport processes in saturated and unsaturated porous media. EPRI Rep. EA - 4190. Electric Power Res. Inst., Palo Alto, Calif.

Gelhar, L.W.; and C. L. Aness. 1983. Three-dimensional stochastic analysis of macrodispersion in a stratified aquifer. Water Resour. Res. (19), 161 - 180.

Ghanem, R.G. and P. D. Spanos. 1991. Stochastic finite elements: a spectral approach. Springer-Verlag, New York.

Gray, W.G. 1975. A derivation of the equations for multi-phase transport. Chem. Engng Sci., (30), 229 - 33.

Gray, W.G.; A. Leijnse; R. L. Kolar; and C. A. Blaiu. 1993. Mathematical tools of changing spatial scales in the analysis of physical systems. C.R.C. Press, FL, USA.

Hassanizadeh, S.M.; and W. G. Gray. 1979. General conservation equations for multi-phase systems: Averaging procedures. Adv. Water Resour., (2), 131 - 44.

Hersh, R. and R. J. Griego. 1969. Brownian motion and potential theory. Scient. Am. (220) 66 – 74.

Kakutani, S. 1945. Proc. Jap. Acad. (21) 227 – 233.

Kloeden, P.E.; and E. Platen. 1992. Numerical solutions of stochastic differential equations. Springer-Verlag, New York.

Klebaner, F. C. 1998. Introduction to stochastic calculus with applications. Imperial College Press.

Knight, F.B. 1981. Essentials of Brownian motion. American Math. Soc.

Kumar, P.; T. E. Unny; and K. Ponnambalam. 1991. Stochastic partial differential equations in groundwater hydrology. Part 2. Stochastic Hydrol. Hydraul. (5), 239 – 251.

Kutoyants, Y. A. 1984. Parameter estimation for stochastic processes. Herderman Verlag.

Lallemand-Barres, P. and P. Peaudecerf. 1978. Bulletin, Bureau de Recherches Géologiques et Miniéres, (3/4), 277-284.

Mathematica Version 3.0, 4.0. Wolfram Research, Inc., IL 61820, USA.

Moroni, M. and J. H. Cushman. 2001. Three-dimensional particle tracking velocimetry studies of the transition from pore dispersion to Fickian dispersion for homogeneous porous media. Water Resources Research. (37) 873-884.

Morse, P.M and H. Feshbach. 1953. Methods of theoretical physics, Part1, McGrawHill.

Øksendal, B. 1998. Stochastic differential equations. Springer-Verlag.

Pickens, S.F. and G. E. Grisak. 1981. Water resources research (17), 1191-1211.

Polya, G. 1921. Mathematische annalen (84), 149-160.

Press, W.H., S.A. Teukolsky, W.T. Vetterling; B. F. Flannery. 1992. Numerical recipes in C, Cambridge University Press.

Rashidi, M.; L. Peurrung.; A. F. B. Thompson; and T. J. Kulp. 1996. Experimental analysis of pore-scale flow and transport in porous media. Adv. Water Resour., (19), 163 - 180.

Sawaragi, Y.; T. Soeda; and S. Omatu. 1978. Modelling, estimation and their applications for distributed parameter systems. Springer-Verlag, New York.

Serrano, S.E. 1988. General solution to random advective-dispersive equation in porous media. Stochastic Hydrol. Hydraul. (2), 79 - 98.

Slitcher, C.S. 1905. Field measurements of the rate of movement of underground waters. Water Sup. Paper No. 140, U.S. Geol. Surv.

Sudicky, S.E.A. 1986. Water Resources Research (22), 2069-2082

Taylor, G.I. 1953. Dispersion of soluble matter in solvent flowing slowly through a tube. Proc. Roy. Soc. (London) A219, 186 - 203.

Thompson, A.F.B.; and W. G. Gray. 1986. A second-order approach for the modeling of dispersive transport in porous media, 1. Theoretical development. Water Resource Res., 22(5) 591-600.

Unny, T.E. Stochastic partial differential equations in groundwater hydrology. Part 1. Stochastic Hydrol. Hydraul. (3), 135 - 153.

Whitaker, S. 1967. Diffusion and dispersion in porous media. AIChE J., (13), 420 - 429.

Weast, R.C. 1972. Table F-47, Handbook of Chemistry and Physics, 53 rd edition, The Chemical Rubber Company, Cleveland, USA

Wiest, R.J.M. 1969. Fundamental Principles in groundwater flow. IN: Flow through porous media. Academic Press, New York.

Young, N. 1988. An introduction to Hilbert space. Cambridge University Press, Cambridge.

Index

www.ingramcontent.com/pod-product-compliance
Ingram Content Group UK Ltd.
Pitfield, Milton Keynes, MK11 3LW, UK
UKHW020805290926
15470UKWH00003B/970

9 780444 51102